W9-AGL-261

FORESIGHT AND UNDERSTANDING

NUFFIELD FOUNDATION UNIT FOR
THE HISTORY OF IDEAS

THE ANCESTRY OF SCIENCE (in four volumes)
The Fabric of the Heavens
The Architecture of Matter
The Flux of Nature
Science and its Environment

★

HISTORY OF SCIENTIFIC IDEAS:
A Teacher's Guide

FORESIGHT AND UNDERSTANDING:
an enquiry into the aims of Science

THE GROWTH OF SCIENTIFIC PHYSIOLOGY

THE FIRST MATHEMATICAL PHYSICISTS:
Eudoxus and his Circle

FORESIGHT AND UNDERSTANDING

An enquiry into the aims of Science

★

STEPHEN TOULMIN

Foreword by Jacques Barzun

GREENWOOD PRESS, PUBLISHERS
WESTPORT, CONNECTICUT

Library of Congress Cataloging in Publication Data

Toulmin, Stephen Edelston.
Foresight and understanding.

Reprint. Originally published: Bloomington : Indiana University Press 1961.
1. Science--Philosophy. I. Title.
[Q175.T64 1981] 501 81-13446
ISBN 0-313-23345-4 (lib. bdg.) AACR2

© by Stephen Toulmin 1961.

Reprinted with the permission of Indiana University Press.

Reprinted in 1981 by Greenwood Press,
A division of Congressional Information Service, Inc.
88 Post Road West, Westport, Connecticut 06881

Printed in the United States of America

10 9 8 7 6 5 4 3 2

Prefatory Note

This book originated in the thirty-fourth series of Mahlon Powell lectures, delivered at Indiana University in March 1960. I have rewritten and considerably expanded the lectures, though without substantially changing their scope or argument. My philosophical colleagues at Princeton and elsewhere will recognize in parts of Chapter 2 the material delivered to them under the title 'Prediction and Explanation': parts of the remaining chapters have also seen the light of day in more inchoate forms.

I intend to treat more adequately elsewhere some of the historical examples which are here presented without sufficient documentation. Many of the missing references will be supplied in the *Ancestry of Science* series (to be published by Hutchinson in London and Harper in New York), of which the first volume has already appeared. Much of the historical material is familiar and well established: my debts to the working historians of science are so obvious as not to require detailed acknowledgment. The only case in which my historical enquiries have led me to make markedly original (and therefore questionable) claims is that discussed in Chapter 4. I hope in due course to back up my claims in a projected book on *The Heritage of the Stoics*—a topic on which fascinating light is already being thrown by the work of Dr. C. C. Gillispie and Professor S. Sambursky.

I am grateful to the President and Trustees of Indiana University for the opportunity of preparing and delivering the Powell lectures. In the course of a year as Visiting John Dewey Professor of Philosophy at Columbia University, also, I have been given an invaluable opportunity to concentrate on research and writing. This book is, thus, one of the fruits of a welcome interval in the United States. I have been greatly assisted in its preparation by Miss Nancy Gordon, who typed and re-typed successive versions with exemplary speed and willingness.

STEPHEN TOULMIN

Columbia University
New York
1960

Contents

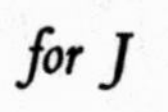

for J

Foreword

JACQUES BARZUN

VERY recently—within the last two or three years—the public decided that it wanted to know 'all about' science. The publishers, docile creatures, have therefore brought out books by the barrelful, books ranging from the layman's encyclopedia in four volumes (a solid and good-looking dust-gatherer for the gentleman's shelf) to the paperback on Magnets (sixty-five cents—a real bargain, but what do I do with this valuable information?). And from other quarters we have had equally numerous discussions of science in relation to our culture, such as C. P. Snow's on the intellectual cleavage between scientists and everybody else, and Bertrand Russell's on the moral issues raised by the kind of work that modern science sets its hand to: no longer the improvement of man's understanding or man's comfort but the increasingly confident assurance of his self-destruction.

Around these curiosities and contentions, usually ill-defined, seldom consistently pursued, stretches the large and dark domain of public ignorance. We all dwell within it. True, a good many people know something of science: they take 'an intelligent interest' and read about atoms and ocean cores and chromosomes and the craters on the moon. Some are practising scientists: they toil at discovery in their subject, following the method and the literature of their specialty with an exclusive, all-absorbing devotion. Of the rest of science they have not the time to learn much. They do not even try, being sure that the only knowledge worth having is that of the professional, who adds new knowledge.

Meantime, the schools enforce upon all above the elementary grades a 'science requirement', which turns out to be largely wasted on three-quarters of those subjected to it. They leave school and college remembering of science only tedium and difficulty. The sole advantage of the compulsion to study an elementary science or two is that it starts the young would-be scientist on his way and perhaps entices a few recruits who did not earlier think of themselves as fit for this career.

In short, Western society today may be said to harbour science like a foreign god, powerful and mysterious. Our lives are changed by its handiwork, but the population of the West is as far from understanding the nature of this strange power as a remote peasant of the Middle Ages may have been from understanding the theology of Thomas Aquinas. What is worse, the gap is visibly greater now than it was a hundred years ago, when educated men could master the main conclusions and simple principles that governed physics, chemistry, and biology. The difficulty today is not that science has uncovered more facts than one mind can retain, but that science has ceased to be, even to scientists, a set of principles and an object of contemplation.

Do we conclude, then, that the situation is hopeless? Do we accept the prospect of the sciences subdividing indefinitely, each specialty becoming the possession of a few workers, while the public stands outside, gaping at the jargon and, once in a while, at the practical results?

Some observers, among whom the author of this little book was one of the earliest, believe that there is a way out of ignorance and into a better light. They base themselves on the experience of mankind in other realms of thought and argue that the public (as against the professional) understanding of such subjects as art, ethics, international relations, ancient and modern history, does not depend on being a performer in these fields. One does not have to be an architect to judge intelligently of houses and monuments; to be a politician in order to have sound opinions about world affairs; to be a religious reformer in order to think about morality; to be a scholarly researcher in order to grasp the history of one's country. What *is* required is that one learn a sufficiency of facts and principles, including the principle that governs the particular enterprise.

To say this is to say that all the subjects in which the thinking public takes an interest are in effect treated philosophically and historically. The article that informs you about modern painting tells you who did what, when, and according to what theory or intention. The outlook upon the admired product is not the technical outlook of the producer but the critical outlook of the appreciator.

The same possibility exists for science to be judged and appreciated—indeed to be enjoyed. And this possibility will have to be realized, with the aid of competent interpreters, if science is ever to become a part of the public consciousness in the same sense as art, history, religion, and philosophy.

It is this work of interpretation that Professor Toulmin has undertaken in a variety of ways, of which these chapters are one. The setting forth of facts and ideas required for a first understanding of science must necessarily take many forms, each adapted to the needs of a particular audience. Children must be taught the elements and shown films, the adult public lectured at and given suitable readings, the scientists gently led to see that the great synthesis towards which they are working will not make itself, but will require the effort of the most self-aware among them.

Every group, moreover, needs to be persuaded that to learn science historically does not mean going back to the lore of the Babylonian shepherds or satisfying a merely antiquarian curiosity. It is modern history, history since 1500, that holds the clue to the characteristic forms and effects of scientific thought. Precisely because science is a stupendous achievement of the human mind, its worth and greatness fully appear only when its successive steps are reviewed with close and informed attention to circumstance, that is, once again, critically and historically.

To know what science is, what it does, and how it affects other manifestations of mind is a task for the man who is at once critic, historian, and philosopher, and who has also been trained in one of the sciences as well as mathematics. Professor Toulmin, who is qualified in these ways, has the added merit of being a lucid and lively writer. Whether one agrees or not with his strongly reasoned conclusions, one feels on reading him that he is advancing the cause of understanding. His book in fact is open to only one objection, which is that it is too short.

I

Introduction

ANY activity has two aspects. One may consider a physical activity like tennis-playing, a practical one like nursing, or an intellectual one like theoretical physics; and in each case the aspect seen by the outsider differs in important respects from that which engrosses the professional. Only the practitioner can understand the training and practice, discipline and method, strategy and imagination called for in the supreme execution of his activity. Yet, at the same time, he may be so close to the activity that its most general features and widest connections begin to escape him. As a result, when it becomes necessary to stand back and appraise the whole sport or profession or intellectual discipline, the outsider once again has something to contribute—not insight (one might say) but 'outsight'.

The present enquiry attempts something very hard—to focus on science something of the insider's judgment and the outsider's breadth of vision alike. In the course of his work a professional scientist generally commits himself to some particular line of enquiry: the great advances in scientific understanding during the last three centuries have, in fact, been largely owing to this division of labour. Still, certain general questions about science remain which, though not immediately urgent for the working scientist, are none the less worth asking. In dealing with these questions (as we shall see) the scientist must co-operate with the historian and with the philosopher.

Our central concern here will be with questions of just this sort. We must try to characterize the *aims* of science: to give, that is, some account of the distinctive purposes and goals properly pursued in scientific enquiry. This is a task quite unlike, and more general than, stating the aims of single scientific investigations. Ask any scientist his aims, and he will indicate those particular aspects of Nature he is trying to explain: whatever he may be—crystallographer, plant bio-chemist, astronomer, or ethologist—he takes for granted that we understand the universal explanatory tasks of the science, and draws attention only to those features distinctive of his own investigations. Yet here we must stand back, and for once ask the wider questions: What is explanation? What makes an investigation or theory a scientific one? These questions are—I shall argue—of central importance both for the Philosophy of Science and for the History of Science.

Our problem has two faces. From one point of view, it calls only for labelling or classification. 'What makes an activity "scientific" rather than (say) "sporting" or "artistic"?': to answer this, we must state the characteristic features by which we classify a theory, or idea, or investigation as a 'scientific' one—so admitting it into the category of science. Looked at this way, the question becomes one of simple intellectual taxonomy. Yet it has another face also, concerned, not with classification, but with *appraisal*. The aims and purposes by which we mark off science taxonomically from sports, politics, and so on, also imply standards for judging the scientist's achievements. So—by implication—our central question asks, further, what makes a scientific theory, or idea, or investigation a successful or unsuccessful one? Quite clearly, this is a matter of evaluation; for answering it means explaining the things that constitute strong or weak points in a new scientific idea. In a phrase, we must state the criteria of *scientific merit*. To give an exhaustive account of the aims and purposes of science means accepting certain general standards of judgment. Good science fulfils these purposes: bad science does not.

It is, in fact, doubtful whether a *final* account could ever be given of the aims of science: especially one which was both exhaustive and brief. Philosophers are often tempted to offer portmanteau characterizations of science, finding in some one requirement (such as predictive success) the unique test of a scientific hypothesis. Our first task here will be to show why one cannot hope to get any real understanding from such a nutshell answer. There is no universal recipe for all science and all scientists, any more than there is for all cakes and all cooks. There is much in science which cannot be created according to set rules and methods at all. And (as we shall see) even the *general* nature of science itself is something in a state of development. Our standards of judgment are liable to amendment, and vary from one field of study to another; and, in some cases, it actually happens that a strong point in one theory turns out—in a different context—to be a weak point in another.

A nutshell definition of science—as of anything else—inevitably floats around on the surface. An investigation of any depth forces us to recognize that the truth is much more complex. To understand the ways in which meritorious scientific ideas differ, in any age, from their less deserving rivals calls for a painstaking and laborious study: only in this way shall we bring to light the manifold functions that science has performed, performs now, and *might* perform in the future within our whole intellectual economy.

Here the historian and the philosopher must reinforce

the scientist. The investigation requires a double-pronged attack—a 'pincer' movement. The critical questions which a philosopher brings to science need to be co-ordinated with the factual studies of history. We must be prepared to scrutinize in detail a representative selection of classic theories, analysing critically the merits for which they came to be accepted, and the standards by appeal to which they established their claims to superiority. The theories in question need not, of course, be dead and gone. In the evolution of scientific ideas, as in political history, the critical analysis of contemporary developments is quite as illuminating as the post-mortem dissection of completed episodes.

In one small book there is room to give only the lineaments of an answer to our central question; but this should be enough to replace the first over-simplified picture of science by the beginnings of a more truthful and lifelike one. Our discussion will have three phases. The first need is to exorcize the dream of stating the central aim of science in a single, all-embracing phrase. (Words like 'prediction', as we shall see, conceal hidden ambiguities. Science is certainly not a matter of forecasting alone, since we have to discover also explanatory connections between the happenings we predict.) Our second and chief business is to examine some selected examples which illustrate what scientific explanations involve in practice. (We shall be forced at this stage to recognize the importance of certain 'ideals of natural order' and 'explanatory paradigms', closely related to Collingwood's 'absolute presuppositions', which have established themselves and developed in the course of men's intellectual history.) Finally, we shall come to see that there is *one* analogy in terms of which the development of scientific ideas can be made immediately intelligible without gross over-simplification. In evolutionary biology, the 'survival-value' of a species needs to be related both to its environment and to its ancestry. And the problem of 'scientific merit' will turn out to be a similar one: it is the problem of seeing in how many ways a novel scientific idea may, in the conditions of its introduction, be 'better adapted' than its predecessors or rivals.

Science has not one aim but many, and its development has passed through many contrasted stages. It is therefore fruitless to look for a single, all-purpose 'scientific method': the growth and evolution of scientific ideas depends on no one method, and will always call for a broad range of different enquiries. Science as a whole—the activity, its aims, its methods and ideas—evolves by variation and selection. Our subject will be the merits which this evolution requires.

2

Forecasting and Understanding

DEFINITIONS are like belts. The shorter they are, the more elastic they need to be. A short belt reveals nothing about its wearer: by stretching, it can be made to fit almost anybody. And a short definition, applied to a heterogeneous set of examples, has to be expanded and contracted, qualified and reinterpreted, before it will fit every case. Yet the hope of hitting on some definition which is at one and the same time satisfactory and brief dies hard: much can be learned by seeing just how much elasticity is ultimately required of such portmanteau definitions.

So in the present case: one may reasonably start by hoping to characterize the programme of science and the nature of scientific explanation *briefly*, in terms of a single aim. To answer our central question by saying, simply: 'The purpose of science is to discover reality', or: 'The purpose of science is to predict correctly', would save us a lot of work. But this gives rise to a prior question. To begin with, therefore, we must ask: 'Has Science a single purpose and aim, or is it rather a multi-purpose activity?' And as an introductory analogy, so as to get a lead, let us begin from the corresponding question: 'What is the purpose of Sport?'

Faced with this question, we may at first answer briskly: 'The purpose of Sport—of a game—is to beat one's opponent: to score more goals and points than him, and in doing so to defeat him.' So far as it goes, this answer is unobjectionable. Unfortunately it does not take us very far. For there are games in which one has no opponent, such as patience or solitaire: a man who plays solitaire is not playing against anybody, so the purpose of that kind of game cannot be to defeat one's opponent. Our initial definition is not completely general: the most one can say along these lines is that the purpose of *competitive* games is to defeat one's opponent.

Behind this seemingly trivial point there is a more serious difficulty. For, if we are told that the purpose of competitive games is to beat one's opponent—that competitive games are competitive, in fact—just how much has one actually been told? This statement turns out to be devoid of substance, and gives us only the bare form of an answer. But our initial question called for something better than a tautology: in order to understand the purposes of sports we must go below this superficial level and see what the phrases 'beat one's opponent' and 'score more than him' amount to in real life.

One thing immediately becomes apparent. In any particular competitive game with an inner variety and structure, e.g. tennis, there are many ways in which points can be scored. Matches can be won, victories gained, as a result of countless different tactics and strategies; and any of a great range of things can qualify you for points. In appropriate circumstances you can add to your score without doing anything yourself—for instance, if your opponent double-faults—and, in exceptional cases, you may even be declared the victor without having so much as raised a finger—if your opponent declines to serve his opening service it will be the umpire's duty to declare him the loser. So, behind simple portmanteau-phrases like 'scoring points' and 'beating your opponent', there lies a complex story. Winning holes in golf is in some ways like going through hoops in croquet, but winning a complete game of croquet is much less like winning a complete round of golf; and neither of them is at all like tennis, or baseball—to say nothing of cricket. To understand the marks of merit and criteria of success on the sporting field, we must explicate and expand our first nutshell definition, until the story which it tells becomes at once less clear-cut and more life-like.

By the time we can give a satisfactory characterization of the nature and aims of sporting activities, we shall be forced to abandon the original idea that sports have a *single* aim and purpose. So long as we rely on a portmanteau answer this idea may appear innocent enough; yet if we want an answer with real substance its implications are unacceptable. Competitive games are of course competitive, and this fact determines the general *sorts* of purposes they will have. Yet in playing any particular game a man will be trying to do a large number of different things—to ace his opponent, tire him out, get him out of position, and so on—pursuing a range of different aims, any of which may contribute to his overall success. It will be his business to add to his score of points or goals in whatever way will best promote the general strategy of his activity. And we shall understand the game in question only when we understand, in outline at any rate, this whole plural range of aims and purposes which a participant in it has to pursue.

. . . . .

What is true of Sport is true of Science also. Each term covers a range of activities, similar in many respects, dissimilar in others, and it proves equally difficult in both cases to encompass the whole range of activities in a single phrase.

Suppose the Philosopher of Science begins by asking: 'What is the (singular) purpose of Science?' We may at once turn his question by replying: 'Has Science *one* purpose?' As our first analogy suggested, the intellectual and practical activities of scientists, like the sporting activities of game-players, have a great range and variety of purposes, which can be only misleadingly summed up in a nutshell definition.

This point is worth demonstrating beyond dispute. So let us suppose that we ask: 'What makes one scientific theory, system, or hypothesis better or worse than another?' As before, the first temptation is to give a simple, clear-cut, and brief reply: 'The better theory, idea, system, or hypothesis is the one that explains more.' And from this point on, the dialectic continues as before. We begin by noticing that this answer cannot be completely general, since not all sciences even attempt to give explanations. Some biological sciences are, rather, therapeutic or classificatory, and succeed in treating or classifying things which they scarcely pretend to explain. Diagnostic medicine (for instance) certainly overlaps into the field of science, and taxonomy is an essential part of botany and zoology, yet in neither is the giving of explanations the necessary purpose of the scientist's activity. And, if the goal of some branch of science does *not* consist in explaining, it can scarcely be criticized on the count of *failing* to do so.

Once again we may be tempted to dismiss these counter-examples as untypical. Such ancillary activities (we may reply) acquire the title of 'science' only from association with the fundamental scientific activities, whose task *is* to 'explain'. Yet this reply, in effect, limits the scope of the term 'science' in an arbitrary way, and prejudices the answer to our own question in the process. For, after this step, our original answer looks a good deal less substantial, having now become:

'A better explanatory theory is one that explains more than a worse one.' Once more we are landed with a tautology: the purpose of competitive games is competitive and the purpose of explanatory sciences is explanatory.

To get more than the bare form of an answer to our question, we must look below the surface and see what 'explaining' involves in practice. The term is, as it stands, an entirely general one, which needs unwrapping in particular detail, in the way the phrase 'scoring points' needed unwrapping. Just as one cannot hope to understand (e.g.) what an 'ace service' is in tennis, without acquiring at least a good spectator's knowledge of the game, so we can hardly hope to show what characterizes a master-stroke in (say) atomic physics, unless we acquire an outsider's understanding of the aims and strategies of that science. And how far these aims and strategies will still apply as we move from atomic physics to biochemistry, or meteorology, or animal psychology, is an open question.

However, before we launch ourselves irrevocably into an empirical study of the methods of the different natural sciences, we must face one last question. 'Is there not *one* thing common to the different explanatory sciences, which unites all sciences more closely than all games? Are not explanations essentially the means of making *predictions*? One need not get bogged down in tautologies, for a perfectly good and clear account can be given of what explanation entails. The purpose of an explanatory science is to explain —that is, to lead to predictions; and the merits of a scientific theory are in proportion to the correct predictions which it implies.'

This view of science has, in recent years, been very popular among philosophers, and it would be silly to ignore it. (If I now criticize it vehemently, that is partly because I once held the view myself.) In scrutinizing it, we must bear one question particularly in mind. Replacing the terms 'explanation' and 'explanatory power' by 'prediction' and 'predictive success' will help us, only if doing so takes us on to a new level—one at which we begin to recognize, in real-life terms, the force and structure of explanation and explanatory power. There is no merit in trading one word for another equivalent one. We want to grasp the substance behind the words.

To anticipate: the predictivist account of explanation fails to help us in this. At first the statement 'A successful explanation is one that yields many predictions' appears genuinely informative and illuminating. But it proves to be so only so long as we leave certain ambiguities in the key term unresolved. For the word 'prediction' is in fact a very slippery one. It slides between two extreme uses: one naive, the other sophisticated. In its most obvious and appealing sense, explaining and predicting are emphatically *not* all-of-a-piece; but, by hedging the term around with sufficient qualifications, we *can* at last use it to provide a definition of explanation. Unfortunately, the effect of all the hedging and qualification is to leave our original problems entirely unsolved. There proves, in the end, to be no substitute for a direct and detailed enquiry into the nature of explanation itself. And how we should embark on this enquiry begins to become clear in the light of the very contrasts between scientific 'explanation' and simple, naive 'prediction'.

. . . . .

In philosophy, as in the law-courts, words which are not defined explicitly must at the outset be interpreted in their current vernacular signification. So here: the terms 'predict', 'prediction', and 'predictive' can most clearly be understood in their familiar, non-philosophical sense. (Incidentally, this is the *only* sense which the dictionaries acknowledge.) On this straightforward interpretation the 'predictivist' account of explanation claims the following: 'When we talk of the "explanatory power" of a theory, we mean its predictive success. And what are predictions? Why: they are pre-dictions, fore-tellings, sayings-in-advance. The crucial test of a theory's power is the number of successful forecasts to which it leads.' We may call this the *First Predictivist Thesis.*

According to this nice, tidy doctrine, the logical character of science is entirely un-mysterious, and it holds out to us the promise of a simple, sure, and even quantifiable test for choosing between good theories and bad ones. As a bonus, the thesis has one particular attraction for professional philosophers: all those references to 'counter-factual conditionals' and 'natural necessity', which have aroused so much philosophical perplexity, are seemingly superfluous. The proof of a pudding is in the eating, and the proof of a theory is in the predicting. Henceforward, we can say all we need about explanation, without going beyond the range of decent, categorical, verifiable forecasts.

How can we test the soundness of this First Thesis? The only way is by measuring it against some typical instances, and in doing so we soon hit on a *First Counter-Argument.*

The thesis defines explanatory power in terms of forecasts. From this it follows that a theory leading to no forecasts is without explanatory merit, and can barely qualify for the title of 'scientific'. Yet plenty of powerful theories have led to no categorical, verifiable forecasts whatever. One obvious example is Darwin's theory, explaining the origin of species by reference to variation and natural selection. No scientist has ever used this theory to foretell the coming-into-existence of creatures of a novel species, still less verified his forecast.

Yet many competent scicntists have accepted Darwin's theory as having great explanatory power. So, whatever they took for 'explanatory power', it could not have been power to forecast those events which the theory explained.

Is this counter-argument too rigorous? Perhaps it can be met by qualifying the original thesis. A *First Qualification* might run: 'For an explanation to be powerful, the theory relied on need not be required to forecast events of *all* the sorts it can explain. Provided it enables us to foretell smaller-scale happenings of the same general kind as the larger-scale ones we use it to explain, that is enough. We can certainly discuss rival theories about the origin of the solar system, without having to bring a fresh solar system into existence or wait for one to be born of itself. The step from forecasts to explanations merely extrapolates from short-term, small-scale events that we can forecast, to similar, larger-scale and longer-term ones that we cannot. So with Darwin's theory: when Australians used myxomatosis to control the rabbit population, it was forecast on the best Darwinian principles that a new strain of rabbits would become dominant, whose constitutions were more resistant to the disease than the average members of the original population. The correctness of this prediction has helped to confirm the merits of the Darwinian theory. And the same thing has happened in other small-scale cases involving melanism in moths, the reactions of infective micro-organisms to antibiotics, and so on.'

Let us again try the Thesis out for size, allowing for the elasticity introduced by this Qualification. As we soon find, the amendment goes only part-way towards meeting the First Counter-Argument. For the history of the Darwinian theory provides a *Second Counter-Argument*. It has, in fact, been possible only in the last few years to make positive, verifiable forecasts on the basis of the theory—even on a small scale. Yet, regarded as a theory, the merits of Darwin's ideas have been actively discussed for a full century; and for most of this time their explanatory power has been unquestioned. Actual forecasting became possible only with the development of modern ecology and genetics, yet men did not wait for this before recognizing the explanatory merits of the theory of natural selection. So it again looks as though the identification of explanatory power with predictive success was premature.

Of course, a *Second Qualification* may do something to rescue the original thesis. 'It is time,' someone may retort, 'to free ourselves from the tyranny of the dictionary, and stop restricting the idea of "prediction" to its naive, vernacular signification. Circumstances no doubt made it difficult for Darwin's contemporaries to forecast future evolutionary changes categorically—human life is too short and biological evolution too slow. Yet it would be churlish to limit the word "prediction" so as to cover only cases in which the events predicted, but not yet confirmed, were themselves still in the future. One can, surely, also predict the nature of things yet to be discovered about the past. The merits of the Darwinian theory, for instance, were confirmed again and again through the guidance it gave to palaeontologists, who brought to light evidence of past phases (e.g. in the development of the horse) which there would otherwise have been no reason to suspect.'

This Second Qualification calls for a *Comment*. The reply may be granted—only its effect is to replace the original Thesis by another. In its ordinary signification, the word 'prediction' certainly does not cover an assertion that some past event *has* taken place. An eclipse or storm or revolution which is already over and done with cannot be *forecast*. The time for that is past. By allowing that Darwin's theory predicted the (future) discovery of unsuspected (past) phases in zoological history, we may gloss over the transition; nevertheless the topic of discussion is now a *Revised Thesis*. Originally, we were free to distinguish pre-diction from retrodiction—saying beforehand that something is going to happen from inferring after the event that it has happened. If the light which Darwin threw on palaeontology is evidence of his theory's 'predictive success', that means interpreting the phrase to cover inferences about events at *any* time—past, present, or future—whether we eventually observe the event itself or only its after-effects. The Revised Thesis accordingly identifies the explanatory power of a theory with its 'predictive success' in a new sense: this now embraces the ability to infer the occurrence of any event in question—whether it has already happened, is happening now, or is going to happen in the future.

Match this modified statement of the predictivist view against further examples, and a *Third Counter-Argument* presents itself. For the most striking examples of predictive technique, in this revised sense, are the mathematical techniques used to predict the times and heights of tides, the motions of heavenly bodies, and so on. Yet (as reflection reminds us) some of the most successful techniques for making such predictions have largely lacked the power to explain the events so forecast, having been worked out by trial-and-error and without any theoretical basis; whereas some respectable theories about the very same natural happenings have been predictively almost entirely fruitless.

To illustrate this contrast most clearly, we must go back to a very early period in science, at which the various aims of science had not yet become associated as we now find them to be. The science of astronomy, for example, seems to have had two independent origins. Between the years 600 and 400 B.C. men both in Ionia and in Babylon were introducing new ways of thinking about the heavens, and contributing to the creation of a scientific astronomy. Yet their contributions were quite different in character. In calculating the times and dates of astronomical events—'saving the phenomena', as it was later called—the Babylonians were masters. Their mathematical command of the celestial phenomena was ahead of the Greeks', at any rate until the capture of Babylon by Alexander the Great; and the first Greek astronomer to equal them was Hipparchos of Rhodes (second century B.C.) who was probably in a position to take advantage of their work. The Babylonian command of the calendar, too, was more exact, while their arithmetical technique for forecasting the first visibility of the new moon and lunar eclipses have no parallel in Greek science of the Golden Age. Yet they achieved all this without (to our knowledge) having any very original ideas about the physical nature of the heavenly bodies. We know that they referred to the major planets by divine names, but the texts so far deciphered offer no evidence of any serious speculation about them.

How then did the Babylonians achieve the results they did? The answer is: they computed the celestial motions in a purely arithmetical way. Like men who prepare tide-tables, or economists working on 'time-series', they analysed each of the celestial motions into a set of independent variables, each changing in a regular, predictable manner. Once this was done, they could calculate the variables separately, and recombine them so as to determine beforehand (or after the event) on which days in a given year the new moon would appear for the first time, and whether at a particular opposition between the Sun and Moon there would be a lunar eclipse.

They succeeded in extending this kind of calculation to the movements of the major planets, and tried with less success to apply it also to earthquakes, plagues of locusts, and other omens. Lunar eclipses proved to occur in a regular and predictable way, but plagues of locusts and earthquakes were intractable. How they explained this fact, we do not know: no theory has been found either about the things they were able to predict, or about those which they were not. Both the successes and failures of their forecasting techniques remained at the time unexplained.

The astronomy of the early Ionians, on the other hand, consisted almost entirely of speculation, theory, and interpretation, and scarcely at all of 'prediction'—either prospective or retrospective. They brought in all kinds of homely analogies: circular tubes full of fire, with small holes through which the fire was visible as stars; the Sun being forced back towards the Equator at the tropics by the cold, dense air nearer the poles; lumps of flaming rock; black, non-luminous, invisible bodies obscuring the light of the Moon; or (more acceptably to modern eyes) the Moon borrowing its light from the Sun and lacking any light of its own. Yet what, in predictive terms, was the outcome of all this speculation? Nothing in it could have enabled Thales to foretell an eclipse of the Sun; this was something which even the Babylonians were never able categorically to forecast. And if one does insist on categorical predictions, whether for the future or for the past, the success of Ionian astronomy was pretty slender. If this had been all that mattered the Babylonians would have left them standing.

What, then, are we to say about this example? Can we square it with the revised statement of the predictivist thesis? We can do so, it seems, only at the cost of saying that the Ionians were far worse scientists than the Babylonians. And this price is too steep a one to pay. The arithmetical command the Babylonians achieved was very admirable; yet, when it came to interpreting the heavenly motions, they showed just how devoid of a theoretical basis their forecasting techniques were. Nobody having a proper conception of the differences between eclipses and earthquakes, plagues of locusts and political disasters, could for a moment suppose that they were all alike predictable by the same kind of arithmetical analysis. The Babylonians acquired great *forecasting-power*, but they conspicuously lacked *understanding*. To discover that events of a certain kind are predictable—even to develop effective techniques for forecasting them—is evidently quite different from having an adequate theory about them, through which they can be understood.

Nowadays, of course, we expect a scientist to combine merits of both kinds. Our own astronomy owes a great deal both to the Babylonians and to the Ionians. One group of men developed the first effective forecasting-techniques, the other taught us the use of the speculative imagination. Only if you are committed beforehand to the view that science has a single, all-embracing purpose are you obliged to praise the one and decry the other. And, even in its revised form, the predictivist thesis has this defect: that it obliges us to dismiss the Ionians' astrophysical speculations as 'unscientific' by comparison with the Babylonians' computational astronomy.

This, however, is not the end of the argument. We began by interpreting the term 'prediction' in its ordinary forward-looking signification; under pressure, we allowed the term to be extended so as to include retrodiction also, thus making the predictivist account of the explanation more elastic; and this process can go further. 'For' (someone may ask at this point) 'are we not still placing too restrictive an interpretation on the use of the key-term?' In this way, the predictivist can introduce a *Second Revised Thesis*, which is at last elastic enough to fit all scientific theories.

How can this be done? In the whole argument up to this point, we have supposed that a 'prediction' must always be a categorical, dated assertion about the occurrence of a particular sort of event—whether in the past, present, or future. 'Yet, why' (it may be asked) 'should all predictions be categorical and dated? No doubt a scientist is *sometimes* in a position to make categorical predictions: e.g. of a total eclipse of the Moon on March 13, 1960. But can we not make hypothetical and conditional predictions, also? Surely we can sometimes predict only with qualifications—that, *if* specified conditions are fulfilled, such-and-such an event will happen: or, alternatively, that a given event happened in the past *if and whenever* such-and-such antecedent conditions held.' Certainly, the extent of our scientific command could not be judged fairly, if we confined ourselves to categorical predictions. Indeed, the whole experimental method for testing our theories depends on our ability to make conditional, rather than categorical, predictions. We predict that, if one does so-and-so, the results should be such-and-such, and then we look and see what in fact happens when we do so-and-so. In its second revised form the predictivist thesis asserts that the merits of a scientific theory are to be judged by reference to its predictive success—where this success now covers conditional, hypothetical predictions as well as categorical ones.

Before we ask directly whether this second revised thesis is adequate to its purpose, a *Second Comment* is in order. It may be put colloquially: 'O.K. If that is how we're to interpret the thesis, that is how we're to interpret the thesis.' The implications of this further gloss on the term 'prediction', which make it so much more elastic than usual, will turn out to be crucial. From the example of the Babylonians, we saw that categorical predictions are relatively unimportant as a test of the explanatory power of a scientific theory, since we may discover how to forecast by simple trial-and-error, without any theoretical understanding of the processes involved; so, from now on, we must concentrate our attention on conditional or hypothetical predictions. Scientific experiments, which count for so much in the establishment of theories, test conditional—not categorical—predictions.

We can see why this further gloss is crucial if we bring our history nearer to the present time. Instead of looking at Babylonian and Ionian astronomy, let us consider their respective descendants—on the one hand, the techniques used for preparing the *Nautical Almanac*; on the other, the explanations a physicist would give of the reasons why these techniques work.

Tidal and astronomical computations today are still carried out as they always were. The records of past events are analysed mathematically into independent, cyclical variables, which can then be re-combined to yield the required predictions. Newton's physical theories certainly help us to understand *why* these techniques work, in a way the Babylonians never understood; but they did not revolutionize the techniques of computation, and calculations based on the traditional methods continued for a long time to be more accurate than those based on the best theoretical principles. The most serious of these discrepancies were eliminated only at the very end of the eighteenth century, as a result of Laplace's *Mécanique Celeste*: others lasted on to provide evidence in favour of the theory of relativity.

At first sight there is a paradox here. No scientific theory has ever provided a more striking advance in our understanding of Nature than Newton's, and one thing it certainly explained was why eclipses can be precisely forecast, in a way that earthquakes cannot. Yet the actual forecasts the theory led to were in many cases incorrect. How, then, did Newton succeed in making our celestial and tidal forecasting-techniques intelligible? What does it mean to say that we now understand why our computational techniques work, in a way the Babylonians did not?

We say this, mainly because we now have a number of general notions and principles which *make sense* of the observed regularities, and in terms of which they all hang together. Think how different Kepler's laws of planetary motion appeared after Newton. Kepler discovered that the orbit of Mars was, as near as he could tell, elliptical. He had some ideas of his own about the forces responsible for the planetary motions, but these provided no compelling reason why the orbit should have just the shape it did. So, the elliptical shape of the orbit was, for him, just a tiresome, obstinate, and arbitrary fact—'one more cartload of dung', which had to be brought into his system, 'as the price for ridding it of a vaster amount of dung'. Newton, by contrast, gave us a whole new set of conceptions, in terms of which Kepler's regularities ceased to appear as arbitrary facts. In the new theory, they all made sense and hung together, granted only a few plausible suppositions. Kepler told us: such-and-such *is in fact* the case. Newton showed us that, if we only supposed so-and-so, then on his principles Kepler's facts must be as they are. ('Freely moving satellites, acted on by a single central inverse-square force only, *must* move in conic sections.') True: if Newton's initial suppositions had been unsound—if, for example, the space between the Sun and planets was not effectively empty—then his explanations would have been quite irrelevant. We should have had to find some other reason for the motions being those Kepler discovered. Actually, his hypotheses were perfectly plausible and made sense, not only of Kepler's laws, but of a great many other things as well. Yet notice: their merits were explanatory rather than predictive. They showed us what must happen *if* certain conditions were fulfilled, not what must happen *unqualifiedly*. They thus drew attention to an intelligible pattern of relationships between apparently unrelated types of happening—the ebb-and-flow of tides, the appearances of comets, the fall of stones, and the motions of the planets. This 'nexus' of regularities and connections was the thing that mattered most for Newton: it determined in his eyes, both what did happen in fact and what would have happened if conditions had been otherwise than they were. It formed a network of natural necessities, holding equally for the actual, and for the unfulfilled conditions.

At the outset of this argument I said that definitions, like belts, could be fitted to any collection of customers, so long as they were sufficiently elastic. Having gone this far in our argument, we can certainly still rescue the predictivist thesis; yet is this worth doing? We started with the thesis that a scientific theory is to be judged by the categorical forecasts to which it leads. Under the pressure of counter-examples, this thesis has been replaced by another one: a theory is now to be judged by the number of factual assertions (past, present, *or* future, categorical *or* hypothetical) which it supports. And it is clear, from the example of Newton's theory, that the word 'supports' here means 'makes sense of' or 'explains'.

We started out to define explanation in terms of prediction; but now prediction itself can be made to meet the case, only if we import into it the idea of 'explaining' and 'making sense of' natural connections. Newton tells us that central inverse-square forces and conic-section orbits necessarily go together: one can, on his principles, infer the one from the other. No doubt we may call inferences of this sort 'predictions', if we please, and say (e.g.) that the Newtonian theory *predicts* that the planets will move as they do, given only the supposition of an inverse-square gravitational force and an effectively empty interplanetary space. We *can* call all these inferences predictions (I say) if we choose: but does it matter any more?

The examples we have looked at in the course of this long argument have shown us something of the logical problem which faces us when we ask about the purposes of science. We hoped at the start to find some brief account of the matter, which could be made to fit all cases without becoming tautologous. (The prime attraction of the predictivist thesis was its appearance of being just that.) We hoped, that is, to elucidate what was involved in 'explaining' some natural happening, and the 'explanatory power' of a theory, in terms of some more elementary idea—and 'prediction' appeared to fit the bill. Yet this idea turned out on examination to be a difficult one to handle: it became as flexible as those flamingoes with which Alice tried to play croquet. On the one hand, you can take the term 'prediction' to mean the same as 'explanatory inference'—but then the doctrine that the function of explanatory theory is to yield predictions leads straight back to the original, unhelpful tautology: 'The purpose of explanatory theories is to explain.' On the other hand, you can take the term to mean, simply, 'forecast'—but then it turns out that the predictive success of a theory is only one test of its explanatory power and neither a necessary nor a sufficient one. (Even the epistemological advantages of the predictivist thesis vanished in the course of our argument. What had made it so attractive to philosophers was the prospect of dispensing with 'natural necessities' and 'counter-factual conditionals'. Yet in Newton's dynamics, the classic example of an explanatory theory, these things are back again in full force.) So far as our present enquiry is concerned, all that is fresh and striking about the predictivist thesis turns out to be mistaken, and what is true in it lands us back in a truism. We are left once again with the question: 'What makes a scientific inference an "explanatory" inference?'

. . . . .

Forecasting, then, is a craft or technology, an application of science rather than the kernel of science itself. If a technique of forecasting is successful, that is one more fact, which scientists must try to explain, and may succeed in explaining. Yet a novel and successful theory may lead to no increase in our forecasting skill; while, alternatively, a successful forecasting-technique may remain for centuries without any scientific basis. In the first case, the scientific theory will not *necessarily* be any the worse; and, in the second, the forecasting-technique will not *necessarily* become scientific, just because it works.

If we have a forecasting-technique which not only works, but works for explicable reasons, that is, of course, doubly satisfactory; this will be the case only where we can point to the natural connection or mechanism which accounts for our predictive success. There will probably always remain some forecasts whose success remains unexplained, because no natural connection is known to account for them—for example, the constant conjunction between the lunar cycle and the menstrual cycle. This conjunction serves well enough in practice, for purpose of forecasting; but forecasts of this sort have no rational basis in our ideas about the 'nexes of natural necessities'. Like most amateur weather-forecasts they are unscientific, even when effective.

As an application of science, forecasting is in fact on the same basis as other techniques. Smelting, medicine, animal-breeding: many other crafts besides those of the forecaster have been profitable starting-points for the development of scientific ideas. In each case the crafts began on a purely empirical basis, by trial and error, before their success could be accounted for scientifically. And why should the ability of a theory to explain the craft of forecasting be a better test of its power than its ability to explain why charcoal can be used to promote smelting, or quinine to combat malaria? True, we can do one thing to keep prediction in the picture. We can extend the term 'prediction' still further, so as to cover all crafts, too: e.g. the prediction 'If you add charcoal to a metallic ore before smelting it, the metal will be formed much more quickly.' But once we are forced to call even recipes predictions, the term has surely been excessively diluted.

We can accordingly distinguish scientific predictions and techniques from pre-scientific forecasts and crafts. Any craft may simply be successful as a matter of experience; or alternatively, its efficacy may be intelligible in terms of our general ideas about Nature. To Claudius Ptolemy, eclipse-predicting and the drawing of horoscopes were both empirical crafts, and in the introduction to his *Tetrabiblos* he insisted that both crafts are equally respectable and scientific. Newton's theory explained why Ptolemy's eclipse-predictions had been successful; but it gave us no reason for thinking that a man's personal fortunes could be forecast from astronomical signs. In this way Newton at last provided a substantial reason for distinguishing between astronomical and astrological forecasting. And how did he do this? By showing that the success of Ptolemy's astronomical constructions, too, tied in with his fundamental laws of motion and gravitation. In terms of these ideals of natural order, facts which before appeared only arbitrary came to appear natural and rational. And this recognition, that at the heart of explanatory scientific theories lie 'ideals of natural order', presents us with the central problem for the rest of our present enquiry. Just how (we must ask) do these ideals enter into our explanations, and how can we recognize them for what they are? It is by working in this direction, rather than by stretching the notion of prediction, that we shall begin to answer the questions from which we started. The central aims of science (I shall claim) lie in the field of intellectual creation: other activities—diagnostic, classificatory, industrial, or predictive—are properly called 'scientific' from their connection with the explanatory ideas and ideals which are the heart of natural science.

. . . . .

One further example will serve to introduce the discussion of 'ideals of natural order'. From the argument so far, one might suppose that the predictivist thesis is an abstract philosophical doctrine, entirely unrelated to scientific practice, and debated about solely by philosophers. This is not so. There have been recurrent periods of self-doubt among working scientists also, during which they have questioned seriously whether science can 'explain' anything, or do more than develop calculi intended to forecast phenomena. Indeed, whenever there is a crisis of confidence within any science, some of its practitioners are tempted to draw in their horns, moderate their claims, and reinsure themselves against criticism, with some unassuming remark: e.g. 'After all, we aren't making any claims about reality—we're only trying to find mathematical relations between observables.'

These disputes about method turn up in ancient, mediaeval and modern science alike: there is something of this tone even in Newton's disclaimer that he is not 'feigning hypotheses' about the cause of gravity. Here, let us look at a sixteenth-century incarnation of the debate—as represented in the person of Nicolas Copernicus, the father of modern heliocentric astronomy, and his colleague Osiander who saw the treatise *On The Revolutions of the Heavenly Spheres* through the printing-press.

Some years earlier Copernicus had given a preview of his theory in the privately circulated *Commentariolus*. It is axiomatic (he declared in its first paragraph) that some sorts of natural happenings stand to reason, being self-explanatory, natural, and intelligible of themselves. The task of astronomy (he declared) was complete only when all the celestial motions were displayed as explicable in terms of 'the principle of regularity'. What was this principle of which Copernicus wrote? It laid down that, in Nature, all bodies which are in their proper places move uniformly and regularly—and this, for him, meant that they moved along tracks composed out of circles, each of which was revolving at a constant angular rate about its own proper centre. That (Copernicus said) is what 'the rule of absolute motion requires', and his lifework was a search for geometrical constructions which, by conforming to this ideal, would give a more coherent and consistent picture of the heavens than Ptolemy had given.

Then Osiander came on the scene, charged with getting Copernicus' authoritative work into the hands of the public. Rumours about Copernicus' new system, he knew, had already aroused theological odium, so he added to the book an unsigned preface 'Concerning the Hypotheses of this Work'. In this, he attempted to cover Copernicus against hostile criticism, by asserting that he was attempting to do no more than Ptolemy and his other predecessors had done:

> 'It is the duty of an astronomer [he wrote] to compose the history of the celestial motions through careful and skillful observation. Then turning to the causes of these motions or hypotheses about them, he must conceive and devise, since he cannot in any way attain to the true causes, such hypotheses as, being assumed, enable the motions to be calculated correctly from the principles of geometry, for the future as well as for the past. The present author has performed both these duties excellently. For these hypotheses need not be true or even probable; if they provide a calculus consistent with observations that alone is sufficient.'

In order to drive the point home, he repeated the predictivist thesis quite generally. Calculation was the only thing that mattered:

> 'For it is quite clear that the causes of the apparent unequal motions [of the heavenly bodies] are completely and simply unknown to this art [astronomy]. And if any causes are devised by the imagination, as they indeed are, they are not put forward to convince anyone that they are true, but merely to provide a correct basis for calculation. Now when from time to time there are offered for one and the same motion different hypotheses . . . the astronomer will accept above all others the one which is easiest to grasp.'

The central words, that the aim of astronomy is not to discover 'true or even probable' causes but 'merely to provide a correct basis for calculation', might well have been written by Mach at the beginning of the twentieth century. Osiander provides a classic statement of the predictivist thesis.

Yet this view, though put forward by Copernicus' friend Osiander, was rather the view of Ptolemy. Copernicus knew very well—and had said—that 'the planetary theories of Ptolemy and most other astronomers' were 'consistent with the numerical data'. His objections to Ptolemy did not rest on observational grounds: the difficulties he felt were entirely theoretical ones.

> 'For these theories were not adequate unless certain equants were also conceived; it then appeared that a planet moved with uniform velocity neither on its deferent circle nor about the centre of its epicycle. Hence a system of this sort seemed neither sufficiently absolute nor sufficiently pleasing to the mind.'

Copernicus was aiming to do more than produce a calculus consistent with the observations: Ptolemy had already done that. He wanted to *explain* the celestial motions, by showing that all the apparent anomalies could be accounted for as combinations of certain ideal, 'absolute' motions. These regular motions alone were natural and self-explanatory.

There is one special merit in using this particular example to pose our central problem. Those who build up their sciences around a principle of regularity or ideal of natural order come to accept it as self-explanatory. Just because (on their view) it specifies the way in which things behave of their own nature, if left to themselves, they cease to ask further questions about it. It becomes the starting-point for explaining other things. Yet the correctness of a particular explanatory ideal (as we shall see) can never be self-evident, and has to be demonstrated as we go along. Copernicus recognized that *explanation* in astronomy involves appeal to some principle of regularity or ideal of natural uniformity, and he called this 'the rule of absolute motion'. Yet his own ideal of uniformity was not the same as ours. He felt no need to look for inter-planetary forces in order to explain why the planets follow closed orbits: in his opinion, a uniform circular motion needed no further explanation, and would—in the nature of things—continue to maintain itself indefinitely. Newton's astronomy was in due course to be based on a rectilinear conception of inertia, in which straight-line motion alone was accepted as the entirely natural sort of self-maintaining motion; but Copernicus still accepted rather a rotational idea of inertia, in which a steady spinning motion was self-explanatory.

All the same, by insisting on the need for *some* principle of regularity, Copernicus was very much the modern scientist. This insistence reopened the road towards modern scientific astronomy which Ptolemy had deliberately blocked. So, from now on, I shall set aside the predictivist thesis and concentrate on a different question. For, about any explanatory theory—as contrasted with a simple, predictive calculus—we can always ask what it implies about the Natural Order. There must always be some point in a scientist's explanations where he comes to a stop: beyond this point, if he is pressed to explain further the fundamental basis of his explanation, he can say only that he has reached rock-bottom.

Here, in a man's ideas about the Natural Order, we find out what is in his eyes self-explanatory. The principle of inertia (or whatever the principle is) is something which for him 'stands to reason'. Such models and ideals, principles of regularity and explanatory paradigms, are not always recognized for what they are; differences of opinion about them give rise to some of the profoundest scientific disputes, and changes in them to some of the most important transformations of scientific theory; so a careful look at a few representative examples should be rewarding.

3

Ideals of Natural Order (I)

WHAT is a phenomenon? How do scientists tell when an event has to be recognized as a 'phenomenon'; and how do they know what sort of a phenomenon it is? The predictivist view of explanation distracts our attention from this question, and that is a pity. For it suggests that, when it comes to applying our theories, all events are on a par—in the same way that all tides, sunrises, and eclipses are to the forecaster. If we have a technique for predicting high-tides or eclipses at all, it must apply equally to all such events; and why (one might begin by asking) should it be any different with explanation?

There is, in fact, an important difference here. A prognosticator may forecast all events of a given type equally, but for the scientist a phenomenon is not just *any* event of the sort he is interested in—it is (as the lexicographers rightly say) 'an event . . . whose cause is in question', and particularly one which is 'highly unexpected'. Further, if a phenomenon is an unexpected event, this indicates, not that the scientist neglected or simply failed to predict it, but rather that he had certain prior expectations, which *made* the event unexpected.

So far as the prognosticator is concerned, the course of Nature need consist only of 'one damn thing after another'. He himself is not going to be caught napping, for he has discovered a way of telling what is going to happen next;

but this is not to say that he understands what is happening. The scientist is in a very different position. He *begins* with the conviction that things are not just happening (not even just-happening-regularly) but rather that some fixed set of laws or patterns or mechanisms accounts for Nature's following the course it does, and that his understanding of these should guide his expectations. Furthermore, he has the beginnings of an idea what these laws and mechanisms are, so he does not (and should not) approach Nature devoid of all prejudices and prior beliefs. Rather, he is looking for evidence which will show him how to trim and shape his ideas further, so that they will more adequately fit the Nature with which he wrestles.

This is what makes 'phenomena' important for him. The games-player improves his sporting techniques most quickly by playing against opponents who are just *one* degree his superior. The scientist, likewise, is on the look-out for events which are not yet *quite* intelligible, but which could probably be mastered as a result of some intellectual step which he has power to take. So long as everything proceeds according to his prior expectations, he has no opportunity to improve on his theories. He must look out for deviations that are not yet explained, but promise to be explicable.

'Deviations'—as soon as one begins to characterize phenomena, the very ink in one's pen becomes saturated with revealing words like 'deviation', 'anomaly', and 'irregularity'. All these imply quite clearly that we know of a straight, smooth, regular course of events which would be intelligible and rational and natural in a way that the 'phenomenon' is not. And this is just the conclusion we are now prepared for: the scientist's prior expectations are governed by certain rational ideas or conceptions of the regular order of Nature. Things which happen according to these ideas he finds unmysterious; the cause or explanation of an event comes in question (i.e. it becomes a phenomenon) through seemingly deviating from this regular way; its classification among the different sorts of phenomenon (e.g. 'anomalous refraction') is decided by contrasting it with the regular, intelligible case; and, before the scientist can be satisfied, he must find some way of applying or extending or modifying his prior ideas about Nature so as to bring the deviant event into the fold. Let us now look at some representative cases in which this intellectual procedure is displayed, so as to show something of the function which 'ideals of natural order' have in the development and application of scientific theory.

. . . . .

We may at this stage look back once again into the history of science, this time turning our attention to the seventeenth century. That period saw drastic changes in several branches of science, including two quite fundamental reorientations, which will be our chief topics in this chapter and the next. To begin with, let me illustrate the points I have been making by reference to the internal re-ordering within the science of dynamics, through which Newton's basic conceptions finally displaced those of Aristotle. In the next chapter we shall look at some changes which began seriously only at the end of the seventeenth century, and affected, not the internal organization of one science, but rather the mutual relations between two different sciences—physiology and matter-theory.

In each case a purely chronological account can be given of the experiments and publications and empirical discoveries of the scientists involved; but the intellectual changes which took place in their thought are intelligible only if we go deeper, and attempt to recognize the fundamental patterns of expectation at stake in the disputes. Happenings of sorts which earlier men had accepted as the natural course of events now came (we shall see) to be regarded as complex and anomalous; while others, which had earlier appeared exceptional, anomalous, or even inconceivable, came to be treated as perfect instances of the natural order. But let us get down to the cases.

First, consider the seventeenth-century revolution in dynamics. To bring out clearly the central change this involved, we must begin by looking at the popular caricature of pre-Galilean theories of motion, which can ultimately be traced back to Aristotle. 'Men's ideas about dynamics before Galileo,' this caricature suggests, 'rest upon a simple mistake. Aristotle was a philosopher, or at best a naturalist, rather than a true scientist: he may have been skilled at collecting specimens and miscellaneous information, but he was bad at explaining things; and he put forward certain clearly mistaken views about the ways in which the motion of a body is related to the forces acting on it. The benighted man asserted that the effect of a given force acting continuously upon a given body was to keep it in motion at a constant speed; whereas we have now looked and seen that a constant force produces not a constant speed but a constant acceleration. Aristotle's successors, having an exaggerated idea of his intellectual capacities, trusted to his words rather than to their own eyes, and only the work of that obstinately common-sensical genius Galileo—who refused to allow himself to be befuddled by mere words, and insisted on submitting even the most august and authoritative doctrines to the test of experience—led to this chimaera being blown away into the oblivion where it properly belonged.'

So stated, this may be less a caricature than the caricature of a caricature; though in less blatant forms, or in part, or by implication, one comes across this view often enough. Still, the picture implicit in this account, both of Aristotelian mechanics and of Galileo's own contribution to our thought, embodies a collection of anachronisms and legends exceptional even for the history of science—a subject in which the George Washingtons have for too long been chopping down their fathers' cherry trees. What one must protest against is not only the intrinsic unlikelihood that a man of Aristotle's capacities could have fallen for so elementary a blunder; but even more, the way in which this caricature degrades a fascinating episode into a prosaic one.

What, then, is wrong? To begin with, this picture gives Aristotle credit for attempting to do something he never seems to have envisaged. It treats him as putting forward a mathematical relationship of the sort familiar from modern dynamical theory. The relationship in question could be written either in words, as

Force varies as Weight times Speed

or alternatively in symbolic shorthand, as

$$F \propto W \times V$$

But this can be read into Aristotle's works only through an anachronism. We scarcely encounter this sort of mathematical equation before the sixteenth century A.D.—not just because the notation employed had yet to be developed, but because the very ideas implicit in the use of such equations were worked out only in the years immediately preceding 1600.

Of course, if we accept this equation as an expression of Aristotle's view, and interpret it in modern terms, we shall find it sadly mistaken. For nowadays it would be natural to take the symbol for speed as meaning 'instantaneous velocity', and the symbol for force in its standard Newtonian sense—both of them notions formulated with complete clarity only in 1687. At once objections arise. The term 'weight' now appears entirely out of place, and should presumably be replaced by the term 'mass'; and even so, the ratio of the force acting on a body to its mass surely determines not its velocity but its acceleration. Yet the question ought to be asked: are we taking Aristotle in a sense which he ever intended? If we read things into him, it will not be surprising if we end up by finding him seriously at fault.

How else, then, can Aristotle's thesis be taken? In general, his practice in the *Physics* is to put forward, not precise equations, but at most ratios or proportionalities relating (say) the lengths of time different bodies will take to go the same distances when different degrees of effort are exerted upon them. He presents these examples as concerned with *tasks*: posing his questions in the form: 'If such-and-such a task takes such-and-such a time, how long will such-and-such another task take?'—e.g. if one man can shift a given body 100 yards by himself in one hour, how large a body can two men jointly shift through the same distance in the same time? Aristotle concludes that, within limits, the amount a body can be displaced by a given effort will vary in inverse proportion to the size of the body to be moved; and also, that a given body can be displaced in a set time through a distance directly proportional to the effort available.

Of course (he allows) beyond certain limits this sort of ratio does not apply: a body may be so large that it can be shifted only by a team of men, and will not respond at all to one man working single-handed—he cites the instance of a team of men moving a ship. And he further remarks, with equal truth, that the effect one can achieve by a given effort depends entirely on the resistances to be overcome. A team of men pulling a ship will take longer to go from one point to another across rough ground than to move it the same distance over smoother ground. As a first approximation, and lacking any better definition of 'resistance', Aristotle accordingly puts forward the further proportionality: that the distance travelled in a given time will vary inversely as the strength of the resistance offered to motion.

. . . . .

Three things need saying about these ratios of Aristotle's, before we look at the dynamical innovations of the seventeenth century. The first is this: Aristotle concentrated his attention on the motion of bodies against appreciable resistance, and on the length of time required for a complete change of position from one place to another. For a variety of reasons, he never really tackled the problem of defining 'velocity' in the case when one considers progressively shorter and shorter periods of time—i.e. instantaneous velocity. Nor was he prepared to pay serious attention to the question how bodies would move if all resisting agencies were effectively or completely removed. As things turned out, his hesitations were unfortunate; yet his reasons for hesitating are understandable, and in their way laudable. Though he was a philosopher—and so, in some people's eyes, bound to have had his head in the clouds and his feet off the ground—Aristotle was always unwilling to be drawn into discussing impossible or extreme examples. Leaving aside free fall for the moment as a special case, all the motions we observe going on close around us happen as they do (he saw) through a more-or-less complete balance between two sets of forces: those tending to maintain the motion and those tending to resist it. In real life, too, a body always takes a definite time to go a definite distance. So the question of instantaneous velocity would have struck him as over-abstract; and he felt the same way about the idea of a completely unresisted motion, which he dismissed as unreal. In point of fact (I suppose) he was right. Even in the interstellar void, where the obstacles to the motion of a body are for practical purposes entirely negligible, there do nevertheless remain some minute, if intermittent, resistances.

In the second place: if we pay attention directly to the kinds of motion Aristotle himself thought typical, we shall find that his rough proportionalities retain a respected place even in twentieth-century physics. Interpreted not as rival laws of nature to Newton's, but as generalizations about familiar experience, many of the things he said are entirely true. One can even represent him as having spoken more wisely than he knew. For, where he argued only for rough, qualitative ratios connecting gross measures of distances and time, contemporary physics actually recognizes an exact mathematical equation corresponding closely to them—though, of course, one which relates instantaneous variables of a kind Aristotle himself never employed.

This equation is known as 'Stokes' Law'. It relates the speed at which a body will move when placed in a resisting medium, such as a liquid, to the force acting on it and the thickness (viscosity) of the medium. According to Stokes, the body's speed under those circumstances will be directly proportional to the force moving it, and inversely proportional to the liquid's viscosity. Suppose we take a billiard ball and drop it through liquids of different viscosities in turn—water and honey and mercury: in each case it will accelerate for a moment, and then move steadily down at a limiting (terminal) speed determined by the viscosity of the liquid in question. If the impressed force is doubled, the speed of fall will be doubled: if one liquid is twice as viscous as another, the billiard ball will travel at only half the speed.

The third point combines these two previous ones. The fact is that Aristotle based his analysis on one particular explanatory conception or *paradigm*, which he formulated by considering examples of a standard type; and he used these examples as objects of comparison when trying to understand and explain *any* kind of motion. If you want to understand the motion of a body (in his view) you should think of it as you would think of a horse-and-cart: i.e. you should look for two factors—the external agency (the horse) keeping the body (the cart) in motion, and the resistances (the roughness of the road and the friction of the cart) tending to bring the motion to a stop. Explaining the phenomenon means recognizing that the body is moving at the rate appropriate to an object of its weight, when subjected to just that particular balance of force and resistance. Steady motion under a balance of actions and resistances is the natural thing to expect. Anything which can be shown to exemplify this balance will thereby be explained.

In the case of bodies moving against a sufficiently slight resistance, as we all know, Aristotle's analysis ceases to apply. If you drop a billiard ball through air instead of through water or treacle, it will go on accelerating for a long time: under normal terrestrial conditions, it could never fall far enough to reach the 'terminal velocity' at which Stokes' Law would begin to apply. The factor of paramount importance in this case will for once be the initial period of acceleration, and that was something to which Aristotle paid very little attention. If he had thought more about the problem of acceleration, indeed, he might have seen the need for something more sophisticated than his simple proportionalities.

As things turned out, Strato, the very first of Aristotle's followers to take an active interest in mechanics, turned his attention at once to this very phenomenon. Yet, for many reasons—some of them intellectual, some of them historical—neither he nor his ancient successors made any great progress beyond Aristotle's ratios. It was left to the Oxford mathematicians of the early fourteenth century to add an adequate definition of acceleration to Aristotle's previous accounts of speed, and so to pave the way for the work of Stevin and Galileo and Newton.

• • • • •

So much for the background: what, then, did happen in dynamics during the seventeenth century? Certainly the popular caricature is wrong in one respect: men did not suddenly become aware that Aristotle's views about motion were false, whereas their predecessors had trusted blindly in their truth. Aristotle himself stated his ratios as applying only within certain limits, and John Philoponos (around A.D. 500) made it absolutely clear that projectiles and freely falling bodies could be explained only by bringing in some radically new conception. The problem was, *how* to remedy matters.

In retrospect we can see that the paradigm at the heart of Aristotle's analysis had to be abandoned and replaced by another, which placed proper importance on acceleration. Yet this was not easy: men were accustomed to think of motion as a balance between force and resistance, as much on the basis of everyday experience as through 'blind trust in Aristotle's authority'. They took the necessary steps hesitantly, a bit at a time, and in the face of their inherited commonsense. The most radical single step was taken by Galileo, yet even he stopped short of the conclusion which is generally credited to him.

There is nothing uniquely natural or rational, Galileo

rightly insisted, in a terrestrial body coming to rest when outside forces are removed: rest and uniform motion alike, he argues, are 'natural' for a body on the Earth. Let us only approach gradually towards the extreme case of zero resistance, which Aristotle had denounced as impossible, and we shall recognize this. Think of a ship (say) on a calm sea, and imagine the resistances to motion progressively reduced, until we could neglect them entirely. If that were to happen, said Galileo, the ship would retain its original motion without change. If it had originally been at rest, it would remain at rest until some outside force started it moving; while, if it were originally moving, it would go on travelling along the same course at the same speed until it met an obstacle. Continuous, steady motion could therefore be just as natural and self-explanatory as rest, and outside resistances alone could bring terrestrial bodies to a halt.

By this step, Galileo went a long way towards the classical Newtonian view, but he did not go the whole way. True, he had exchanged Aristotle's paradigm of natural motion—the horse-and-cart being pulled along against resistances at a constant speed—for a very different one. For Aristotle, all continuous terrestrial motion was a 'phenomenon', or departure from the regular order of things, and he would have asked: 'What is to keep Galileo's imaginary ship moving?' Galileo, however, now demanded only that we account for *changes* in the motion of bodies. His ship could move for ever without a motive force.

Now this result looks, at first sight, very like our modern 'law of inertia'. Yet Galileo's paradigm was no more identical with our own than Aristotle's had been. For what he envisaged as his ideal case was a ship moving unflaggingly across the ocean along a Great Circle track, for lack of any external force to speed it up or slow it down. He saw that uniform motion could be quite as natural as rest; but this 'uniform motion' took place along a closed horizontal track circling the centre of the Earth; and Galileo took such circular motion as entirely natural and self-explanatory. He does not seem to have regarded the ship as constrained by its own weight from flying off the Earth on a tangent—the image which can clearly be found in Newton.

Indeed, if Galileo's imagined ship *had* taken off from the sea and disappeared off into space along a Euclidean straight line, he would have been no less surprised—in fact, *more* surprised—than us. We should have one possible hypothesis at hand to explain this amazing event—namely, that the action of gravity on the ship had been suspended, so that it was no longer constrained to remain in contact with the Earth's surface and could fly off along its natural path. For Galileo, however, this option was not yet available: in his eyes, some active force alone could have obliged the ship to travel in a perfectly rectilinear path, instead of cruising of its own accord round its natural Great Circle track.

. . . . .

When we turn to Newton we find that the ideal of natural motion has changed yet again. The fundamental example is completely idealized. From now on, a body's motion is treated as self-explanatory only when it is free from all forces, even including its own weight. Galileo could explain his conception of 'inertia' by referring to real objects—ships moving on the sea. Newton started his theory by offering us a completely abstract example as the paradigm—namely, a body moving at uniform speed in a Euclidean straight line—and this, as Aristotle would have retorted, is the last thing we should ever encounter in the real world. But, then, Newton does not have to claim that, as a matter of fact, any actual body moves exactly as his first law specifies. He is providing us, rather, with a criterion for telling in what respects a body's motion calls for explanation; and what impressed forces we must bring to light if we are to succeed in explaining it. Only if a body ever were left completely to itself would it move steadily along a straight line, and no real body ever actually is placed in this extreme position. This is, for Newton, simply a dynamical ideal, the sole kind of motion which would be self-explanatory, free of all complexity, calling for no further comment—if it ever happened.

It should be clear, by now, why I present Newton's first law of motion or principle of inertia as an 'ideal of natural order'—one of those standards of rationality and intelligibility which (as I see it) lie at the heart of scientific theory. At their deepest point, the seventeenth-century changes in dynamics, which had been brewing ever since the early 1300's, involved the replacement of Aristotle's common-sensical paradigm by Newton's new, idealized one. From some angles, this could look like a regression: from now on it was necessary, for theoretical purposes, to relate familiar everyday happenings to idealized, imaginary states-of-affairs that never in practice occur—ideals to which even the motions of the planets can only approximate. Yet the change paid dividends. Once this new theoretical ideal was accepted, the single hypothesis of universal gravitation brought into an intelligible pattern a dozen classes of happenings, many of which had previously been entirely unexplained; and, in the resulting theory, Newton could display a whole new range of relationships and necessities as part of the intelligible order of Nature.

This example has illustrated how the idea of explanation is tied up with our prior patterns of expectation, which in turn reflect our ideas about the order of Nature. To sum up:

any dynamical theory involves some explicit or implicit reference to a standard case or 'paradigm'. This paradigm specifies the manner in which, in the natural course of events, bodies may be expected to move. By comparing the motion of any actual body with this standard example, we can discover what, if anything, needs to be regarded as a 'phenomenon'. If the motion under examination turns out to be a phenomenon—i.e. 'an event . . . whose cause is in question' as being 'highly unexpected'—the theory must indicate how we are to set about accounting for it. (In Newton's theory, this is the prime task of the second law of motion.) By bringing to light causes of the appropriate kind, e.g. Newtonian 'forces', we may reconcile the phenomenon to the theory; and if this can be done we shall have achieved our 'explanation'. Every step of the procedure—from the initial identification of 'phenomena' requiring explanation to the final decision that our explanation is satisfactory—is governed and directed by the fundamental conceptions of the theory.

No wonder that the replacement of one ideal of natural motion by another represents so profound a change in dynamics. Men who accept different ideals and paradigms have really no common theoretical terms in which to discuss their problems fruitfully. They will not even *have* the same problem: events which are 'phenomena' in one man's eyes will be passed over by the other as 'perfectly natural'. These ideals have something 'absolute' about them, like the 'basic presuppositions' of science about which R. G. Collingwood wrote.

If that is so, the problem at once arises: how do we know which presuppositions to adopt? Certainly, explanatory paradigms and ideals of natural order are not 'true' or 'false', in any naive sense. Rather, they 'take us further (or less far)', and are theoretically more or less 'fruitful'. At a first, everyday level of anaylsis, Aristotle's paradigm of uniform, resisted motion had genuine merits. But a complete mathematical theory of dynamics required a different ideal. It was no good first taking uniform, resisted motion as one's paradigm, and supposing that one could later explain how bodies would move in the absence of resistances by cancelling out the counteracting forces: that way inevitably led to the unhelpful conclusion that a completely unresisted motion was inconceivable—since the attempt to describe it in everyday terms entangles one in contradictions. (Suppose you reduce the resistances finally to zero, then, in Aristotle's ratio of motive force to resistance, the denominator becomes zero; and you are landed in all the difficulties which spring from 'dividing by nought'.) On the contrary: it was necessary to proceed in the opposite direction. One must first start by taking entirely unresisted motion as one's ideal of perfectly simple and natural motion; and only later introduce resistances—showing how, as they are progressively allowed for, the uniform acceleration produced by a single force gives way to the uniform terminal speed of a horse-and-cart.

Changes in our ideals of natural order may sometimes be justifiable, but they do have to be justified positively. In due course uniform rectilinear motion became as natural and self-explanatory to Newton's successors as rest had been for Aristotle. Yet neither view of inertia was self-evidently correct: each must be known by its fruits. So its tenure as the fundamental ideal of dynamics was conditional, and provisional. For just as long as we continue to operate with the fundamental notions of the Newtonian theory, his principle of inertia keeps its place in physics. Yet, at the most refined level of analysis, it has already lost its authority. As one consequence of the twentieth-century changeover to relativity physics, the conception of 'natural motion' expressed in

Newton's first law has again had to be reconsidered. The implications of the resulting amendments in our ideas may have been less drastic than those which flowed from the seventeenth-century revolution; yet—at the theoretical level—the change has been none the less profound.

. . . . .

Before we go on to our second example, let us return to a less rarefied atmosphere. The general point I am making does not apply only to abstract and highly developed sciences, such as dynamics. We use similar patterns of thought in the common affairs of daily life; and, in a sense, the task of science is to extend, improve on, and refine the patterns of expectation we display every day. There is a continual interplay between the two fields.

Suppose, for example, that we look out of the window, into the street. One car travels steadily down the road, comes into sight, passes our window, and goes on out of sight again: it may well escape our attention. Another car comes down the road haltingly, perhaps jerking and backfiring, perhaps only stopping dead and starting up again several times: our attention is immediately arrested, and we begin to ask questions—'Why is it behaving like that?' From this example it is only a step to the case of a practical astronomer, for whom the continued motion round its orbit of the planet Jupiter is no mystery: but for whom questions would immediately arise if the planet were suddenly to fly off along a tangent to its orbit and out into space: 'What made it do that?' And from this it is only one further step to the mathematicians' point of view, according to which, if left to itself, Jupiter ought to travel, not in a closed orbit, but in a straight line—so that even its normal, elliptical path demands explanation.

All the same: though the form of this thought-pattern is preserved, its content changes drastically, and one popular epigram about explanation is falsified in the process. For it is often said that 'explanation' consists in relating things with which we are unfamiliar (and which so need explaining) to others which are familiar to us (and so stand in no need of explanation). At a certain level this epigram has a point. If you are explaining something *to somebody*—what might be called an explanation *ad hominem*—it is sensible to start from things he knows about and understands, and to relate the things he finds mysterious back to those which he finds intelligible. This is one of the purposes of 'models' in the physical sciences. The beginner in electricity is helped to understand the relations between voltage, current, and resistance by having the flow of electricity in a wire compared with the flow of water down a tube: 'Don't you see? Voltage is like the head of water in the system, resistance is like the narrowness of a pipe, and the current of water or electricity depends in each case on both factors.'

Scientific discoveries, however, do not consist in arguments which are plausible *ad hominem*, but rather in explanations which will stand on their own feet. In these explanations, the relation between the 'familiar' and the 'unfamiliar' may be reversed. Revert for a moment to Newtonian dynamics: the ideal of inertial motion which underlies Newtonian explanations can hardly be described as *familiar*. (Aristotle would laugh at that suggestion.) If we were to insist on accounting for the 'unfamiliar' in terms of the 'familiar', instead of *vice versa*, we should never be able to shake ourselves loose of Aristotelian dynamics. Aristotle's paradigm is familiar in a way that Newton's never can be; and the Newtonian programme of treating the motion of horses and carts as being something highly complex, which can be understood only by starting from planets and projectiles and then allowing for a multiplicity of interfering forces—remains rather paradoxical to the commonsense mind.

What are the lessons of this first example? In ordinary life explanation may, perhaps, consist in 'relating the unfamiliar to the familiar'. But, as science develops, this turns into 'relating the anomalous to the accepted', and so in due course into 'relating the phenomena to our paradigms'. This is inevitable. Which things are familiar and which unfamiliar is a relative matter. (A man who lived in a desert might find the idea of 'the head of water' a difficult one to grasp, and be more mystified by hydraulics than by electricity.) On the other hand, whether an event is 'anomalous' or not need not be so personal a question. It can be discussed rationally—still more, if we go to the length of labelling the event as a 'phenomenon' and implying that it needs to be squared with theory. For then our standard must be, not what is familiar, but rather what is intelligible and reasonable in the course of Nature. And where we are led once we recognize this distinction, it has been the aim of this chapter to show.

4

Ideals of Natural Order (II)

NOT so long ago, in the days before it became intellectually respectable, jazz used to be contrasted with 'good music'. Of course (the critics granted) it was not all equally debased: there could be better jazz and worse jazz. For that matter, not all 'good music' was equally good—nor even equally 'good': Grieg, for instance, was a splendid composer of his kind, and yet remained suspect, as not being wholly 'good'. Still, in judging music (it was presumed) one had to consider both whether a piece or composer was a good example of a particular type, and whether that type was a 'good' type. Different *genres* of music were thus placed in a hierarchy, some being subordinated to others.

Looking back, we may doubt whether these cross-type comparisons were fair, or even legitimate. Yet the fact is that they were made. And—to return to our own topic of discussion—a similar pattern of comparison can be found in discussions of science. Just as the question 'Is this music good of its kind?' is distinct from the question 'Is it "good" music?', so we find scientists asking both 'Is this event a natural and self-explanatory one of its kind?' and also 'Is this an example of the most natural and self-explanatory sort?' Explanatory ideals and paradigms thus have two parts to play. Within a given science, such as dynamics, one kind of motion (say) will be accepted as the standard of intelligibility. (The principle of inertia was the instance we examined in the last chapter.) But, when we compare happenings of *different* kinds, we find different sciences being subordinated to one another, and so setting standards for one another, in the same way as *genres* of music. Phenomena may, as a result, be explained either by comparing them with other, more self-explanatory happenings of the same kind or by relating them to happenings of some other sort, which are thought to be intrinsically more natural, acceptable, and self-explanatory. Let us now look at an example of this latter procedure. For, here again, we shall find men's scientific ideas at the mercy of history, and be forced to recognize that the intellectual hierarchy of the sciences has been subject to profound changes.

Let us start with a text. This one is taken from a seventeenth-century essay on chemistry, whose authorship I intend to keep anonymous at the outset.

> 'All Bodies have Particles which do mutually attract one another: the Sums of the least of which may be called Particles of the *first Composition*; and the Collections of Aggregates arising from the Primary Sums, or the Sums of these Sums, may be called Particles of the *second Composition*, etc. Mercury and *Aqua Regis* can pervade those Pores of Gold or Tin, which lye between the Particles of its *last Composition*; but they can't get any further into it; for if any Menstruum could do that, or if the Particles of the first; or perhaps of the second Composition of Gold could be separated; that Metal might be made to become a Fluid, or at least more soft. And if Gold could be brought once to ferment and putrify, it might be turn'd into any other Body whatsoever. And so of Tin, or any other Bodies; as common Nourishment is turn'd into the Bodies of Animals and Vegetables.'

The last two sentences will stand repeating: 'And if Gold could be brought once to ferment and putrify, it might be turn'd into any other Body whatsoever. And so of Tin, or any other Bodies; as common Nourishment is turn'd into the Bodies of Animals and Vegetables.'

To a modern reader, these words—which conclude the whole essay—are strangely at variance with the rest of the passage. If we allow for the archaic terminology—'menstruum' for 'solvent', and so on—it begins by stating clearly the basic principles of an atomic theory of matter. Fundamental particles combine to form atoms, these in turn are joined together into molecules, and so on. The greater the degree of 'composition', the more complex will be the organization of the resulting body. The author thus builds up a picture of material things as ordered structures of fundamental particles. And then suddenly he changes his tone. From an exposition of modern atomism, we switch straight into quaint imaginings: as if gold, for instance, could 'ferment and putrify', and tin be digested like a nutrient and transformed into something else. How (one wants to ask) could anyone who understood the nature of elementary inorganic metals like tin and gold describe ordinary chemical reactions in terms of the metabolic processes by which nutrients are first embodied in the cells of an organism, and eventually released in putrefaction?

The question becomes more insistent when the author of this passage is known. For it comes from an essay on the nature of acids which Sir Isaac Newton wrote in the 1690's, and which was first printed by John Harris in his *Lexicon Technicum* in 1723. Such apparent muddle-headedness might have been passed over in an average seventeenth-century scientist; but was not Newton, that devoted prophet of the 'corpuscular philosophy', granted an almost prophetic foresight of classical atomic physics and chemistry? He is the last man we should expect to find guilty of this particular confusion.

In fact, the questions raised by this passage are well worth facing directly. What precisely is it about this passage that we find unacceptable? How can Newton, even, describe inorganic physical and chemical changes in terms of organic metabolism? We can answer this question adequately only if we set Newton's essay in a much wider context; and, in what follows, we must look at the fundamental structure of explanations in the science of material things—particularly, at the way in which the relations between 'chemical' and 'physiological' ideas have changed in the course of history. To anticipate: the distinction between 'inorganic reactions' and 'organic metabolism', which lies at the heart of our present discomfort, is neither inevitable nor obvious. At the end of the seventeenth century, the lines along which such a distinction was to be drawn could not yet be clear, even to the most forward-looking of men. And between 1600 and 1800 the relative positions of chemistry and physiology were to be completely reversed.

. . . . .

Let us look again at our discomfort. Newton was not the only leading corpuscularean of the time who talked about chemistry in a way that strikes us in retrospect as bizarre. There is, for instance, a well-known letter in which Robert Boyle explains to John Locke why he had never bothered about the problem of alchemical transmutation. This was not through scepticism: he did not question the possibility of turning one metal into another. It was simply (if I may paraphrase him in modern terms) because he had chosen to concentrate on pure science, rather than be distracted into the field of scientific technology. Transmutation could be left to those who wanted to make money rather than theoretical discoveries. He himself had a sufficient private income, and could afford to aim at intellectual light without concern for financial gain: concentrating, in his own words, on 'luciferous' experiments rather than 'lucriferous' ones.

Now, to a straightlaced historian of chemistry, utterances such as these, coming from Newton and Boyle, can be explained only as signs of backsliding or weakness of mind. Surely the atomic theory was so obviously right and scientific that its modern progenitors could have avoided this truckling to the alchemists. Their formulation of the atomic theory appears to modern eyes a turning-point—after which the progress of matter-theory at last becomes cumulative and scientific. What a pity, then, for these great men to lapse into an antiquated and unscientific muddle.

This reaction misses the point. Behind these quotations lies a question which was in fact still unsettled in the late 1600's; and we must bring it to light if we are to see the situation of these men in a proper perspective. This is the question: 'What should be our fundamental type or paradigm of a material change? Should we (say) set about explaining the structure and constitution of things by reference to the manner in which they develop? Or should we, rather, account for their mode of development by relating it back to their material ingredients and structure?' Over this question (as we shall see) opinions have changed radically since 1600, in a way which all modern chemical theory takes for granted.

The intellectual transformation has some similarity to the one we studied in dynamics. Recall the earlier example: how at the outset Aristotle thought of resisted motion at uniform speed as typical, but by Newton's time the theoretical paradigm became the motion of an unresisted, uniformly accelerating body—steady, resisted motions now being regarded as a secondary, un-self-explanatory type. So again in the present case: men's first intellectual model of material change had to be abandoned and replaced by a more sophisticated one. Once again, the men who transformed matter-theory during the eighteenth century stood Aristotle's method of explanation on its head. Where Galileo and Newton between them forged a new conception of 'inertia', chemists in the century or so before 1800 established their science on the basis of an equally novel ideal—that of absolutely 'inert' or 'inanimate' matter. Though one may find in earlier times a more or less clouded recognition that some forms of matter, including the fundamental ingredients of things, are entirely inanimate, this doctrine was never firmly established or universally accepted. Establishing a comprehensive matter-theory on this foundation, in fact, involved a kind of abstract idealization quite as sophisticated as anything in Newton's dynamics. In what follows I shall try to explain and justify this thesis.

. . . . .

Let us start with an elementary illustration. Suppose we set out to compare cooking with ripening. The question arises: is it more illuminating to compare the changes produced by cooking to the process of ripening, or to explain ripening in terms of the effects of cooking? Nowadays, we should probably plump for the latter: we should set about explaining the obvious, visible changes which take place when an unripe ear of wheat turns into a ripe one, by referring to invisibly small physical and chemical changes within the ear.

> 'Ripening? [we might say] Well, you know about cooking—how, under the action of heat in the oven, structural alterations are produced in the tissues of a steak —as a result of which it changes in colour and texture, and becomes easy on the jaw. Ripening is a comparable process. The heat of the Sun, like that of an oven, once again stimulates within the ear of wheat structural modifications whose consequences show up in its colour and texture. To begin with it was green and hard; but now, as a result of these structural changes, it becomes golden and softer.'

The full story is, of course, much more complex than this, and probably has not yet been worked out. For the present purpose, however, the details are not important: it is rather the direction of thought—from physiology to physics and chemistry—with which we are now concerned.

In Aristotle the direction of thought is reversed. He does not set about explaining ripening by comparing it with cooking: rather, he works the other way round. Ripeness is all; material changes tend of themselves towards that goal; and artificial 'concoction' can only speed up the normal, uninhibited processes of Nature.

> 'Cooking? [he might say] Well, you know about ripening—how, as the weeks pass, the seeds germinate, the infant seedlings turn into the adolescent stalks, and the plants finally come to their natural maturity—the innate qualities of the adult plant all the while coming to light, as the process of growth and maturation pursues its natural course. Cooking is a comparable process. The raw steak is no *Filet Mignon* to begin with; but it is capable of developing into one, if subjected to the appropriate

> environmental conditions; for, under these conditions, it is given the opportunity to manifest in fact all the inherent tenderness and succulence of which it is capable.'

For Aristotle, cooking and ripening were both forms of 'concoction', but ripening was the more typical and self-explanatory of the two. In each case the inherent qualities of an 'immature' body were brought out, by subjecting it to the appropriate degree of heat for the appropriate length of time; so that cooking was, so to speak, a kind of artificial ripening. To understand a thing's material nature and constitution meant, in consequence, to recognize what it was capable of developing into, either naturally and of itself, or artificially if appropriately treated. Chemistry was thus subordinated to physiology, instead of the other way round.

Was it ever reasonable to establish a theory of matter on this paradigm of natural change? Looking back, we may feel that the attempt was bound to end in failure. Yet this could hardly have been obvious at the outset. Men were faced with the choice between explaining physiological development in terms of structural change, or structural change in terms of physiological development: either programme might have paid dividends. The modern programme of explaining development in terms of structure was on the whole less commonsensical and more abstract; so it is not surprising that men tried the other direction first. But, whatever the rights and wrongs, we shall never follow the evolution of matter-theory before 1700 unless we do take into account this drastic change in the direction of explanation.

The older programme was the foundation of chemical theory and practice for many centuries, during what we know as the 'alchemical' stage. Alchemy was not just a species of black magic, camouflaged by a pretentious array of jargon.

It was, rather, a premature system of chemical philosophy, founded on a highly developed set of ideas. These ideas embodied and carried further Aristotle's developmental paradigm of material change, and they have left their mark on our language in the form of numerous dead metaphors.

The fundamental axioms of the alchemical philosophy of matter can be summarized quite briefly. We must first put aside our own assumption that the basic ingredients of matter are certain stable, inert, homogeneous stuffs, which retain the same observable properties indefinitely so long as they are not exposed to other substances which react with them. (This is the very paradigm whose adequacy was, in 1600, not yet established.) The alchemists assumed, instead, that all material substances possessed what that great landscape gardener Brown used to call 'capability of improvement'. Nature was in course of self-development; metallic ores, for instance, were pursuing their natural development within the matrix of the rocks—in fact, in what men referred to (quite unmetaphorically) as 'the Womb of the Earth'; and all things in Nature, if left to themselves, would eventually fulfil their inherent potentialities, so far as conditions permitted.

As to the alchemist's technological ambitions: he hoped that, by subjecting materials to the appropriate conditions, he could accelerate the pace of their natural development. In this way, metallic materials could presumably be turned from their immature ('base') forms into their mature and adult ('noble') forms more rapidly than usual. Heat was clearly the crucial controlling factor. The child in the womb and the ear of wheat alike relied for proper development on being subjected to precisely the right degree of heat at the right time. The first principle of alchemical technology was, therefore, to re-create and intensify the quasi-embryological conditions in which material substances were believed to develop. The legendary founder of alchemical technique, Hermes Trismegistos, was credited with inventing a way of producing out of glass an artificial 'womb', which could be perfectly sealed-off by heat—'hermetically' sealed, as we still say. The alchemist then hoped, by suitable concoction within this glass womb, to speed up the growth and maturation of his raw materials.

The idea that veins of metallic ore grow and change by a process of regeneration in the womb of the earth remained popular until the middle of the eighteenth century even in learned circles: one finds it, for instance, in Linnaeus. And the belief that they can 'grow' was not entirely without foundation: mineral substances may be deposited in the earth by water seepage, over long periods of time, very much as stalactites and stalagmites are. What we now reject is the embryological interpretation of this fact—its proper interpretation having become clear only when men began to apply in geology the fundamental insights of Lavoisier and Dalton.

. . . . .

One alchemical idea in particular was to remain influential for a long time. For it was widely believed that, under appropriate conditions, one could hasten the maturation of baser metals by adding a small quantity of a noble metal to the mixture. A few grains of gold dropped—as a 'seed'—into an alloy of silver and copper would in this way accelerate the natural transformation of the alloy into genuine gold. At first sight, this looks like 'salting the mine': rather, the hope was to 'breed' more gold from the baser alloy. And in due course this basic idea, that natural objects are formed through the action of seeds—which had been in circulation since some of the early Greek atomists, e.g. Anaxagoras—gave rise to a complete system of matter-theory.

The best-known exponent of this developmental view of matter was a seventeenth-century Flemish physician, J. B. van Helmont. He attributed all natural changes to the action of certain inner, formative ferments: these were the controlling factors or 'organizers' in things, and they had the power to impose particular properties on the fundamental, characterless raw material—water. But he was not alone in the view. It was, in fact, quite popular around 1650 and appears in the writings of King Charles II's physician, Thomas Shirley. For instance, to account for geological changes, such as the formation of fossils and petrified timbers, he writes:

> 'The Hypothesis is this, viz. That stones, and all sublunary bodies, are made of water, condensed by the power of seeds, which with the assistance of their fermentive Odours, perform these Transmutations upon Matter.
>
> The Seeds of Minerals, and Metals are invisible Beings; (as we have shewed, above, the true Seeds of all other things are;) but to make themselves visible Bodies they do thus: Having gotten themselves suitable Matrices in the Earth, and Rocks (according to the appointment of God, and Nature) they begin to work upon, and Ferment the Water; which it first Transmutes into a Mineral-juice, call'd Bur, or Gur, from whence by degrees it formeth Metals. . . .
>
> The Saxeous, or Rocky Seed, contained in these Waters, (which is so fine, and subtile a Vapour, that it is Invisible; as I have before shewed all true Seeds are) doth penetrate those Bodies which come within the Sphere of its Activity; and by reason of its Subtility, passes through the pores of the Wood, or other Body, to be changed. . . .

So this Stonifying Seed, by its operating Ferment, doth transchange every particle of the matter it is joined unto, into perfect Stone.'

To anyone with a modern scientific education, the system of matter-theory put forward by Shirley and van Helmont sounds highly fanciful. Yet, in one curious respect, we are intellectually in their debt. Notice how, in the passage quoted, Shirley coins two barbarous, monosyllabic neologisms as names for his 'mineral juice', viz. *Bur* and *Gur*. These linguistic inventions mark him off as a true disciple of Helmont, who himself set the fashion by coining, entirely out of his own imagination, the similar terms *Blas* and *Gas*. This was, in fact, the way in which the word 'gas' entered linguistic history. And Helmont did not just invent the word: he was also one of the first men to experiment seriously with gases and to recognize clearly that aeriform (or gaseous) substances are not all chemically identical.

Helmont became convinced that there are many varieties of gas, not so much on experimental grounds as for reasons of theory—*his* theory. As he saw it, objects of all kinds had the same basic raw material (water), but each particular thing had a highly specific form (was a cow, or a willow-tree, or a lump of quartz) through the action of its own formative 'ferment'. Suppose you take an oak-log and burn it: the effect will be to break down the characteristic form and organization created when the 'ferment' in the oak seedling ordered the water drawn in from the roots: only the cremated ashes of the original tree will be left behind. When the form of the oak-tree is destroyed, Helmont argued, the formative ferment must be released: it escaped, he concluded, in the form of a vapour or 'spirit', which cannot be enclosed but may in some cases, owing to its close association with water vapour, be condensed. This aeriform vapour, comprising a vital ferment or 'spirit' in association with water, he christened a 'gas'. And, since different objects have different forms, the gases driven off in combustion must carry off equally different formative 'spirits', and must likewise differ from one another.

The conclusion was right; even though the argument now strikes us as merely quaint. There are in fact many different gases, but not for the reason Helmont gives. Nor do we think, as the alchemically-minded perfumiers used to do, that the creative souls of plants can be driven off in an evaporating vat and distilled again in concentrated form. Yet this was the original implication of the phrase 'essential oils', which is still used by perfumiers today. The active essence in the living lavender-plant (it was supposed) converted rain-water in the soil into the stalk, branches and leaves; and the distiller's task was to boil the leaves in water until the form was destroyed and the vital essence driven off. Once again, this conception has a natural basis: the art of capturing the fragrance of a plant in a lasting form, separate from the dead plant, was a striking achievement—almost as though a vial of perfume did indeed represent 'a soul in a bottle'. Helmont himself comments on the action of perfume:

> 'The Spirit of our life, since it is a Gas, is most mightily and swiftly affected by any other Gas, by reason of their immediate co-touchings. For neither therefore does anything thereupon operate more swiftly on us, than a Gas; as appeares . . . in perfumes. . . . For a Gas is more fully implanted and odours keep a more immediate co-touching with the vital Spirits, than Liquors.'

The sort of process which Helmont and Shirley regarded as the natural paradigm of material change is becoming important once again in the mid-twentieth century—though at a technological, not a theoretical, level. During the last few years a new interest has arisen in 'biochemical engineering'. Bakers and brewers have for centuries used yeasts and ferments, and now a more sophisticated and deliberate exploitation of enzymes and similar substances to control and accelerate biological changes is holding out novel and exciting prospects. Pharmaceutical manufacturers use a process of 'culturing' to multiply penicillin and other antibiotic moulds; and the range of such processes being exploited in commerce is large and increasing. True, it does not include gold-production. Yet the dream of producing gold in this way was never nonsensical—only vain. Indeed, the process of alchemical 'multiplication', by which a few grains of gold were intended to leaven a baser alloy and turn it all to pure gold, probably appeared to the 'adepts' as a kind of 'culturing'. And so it might still do to us if we had not established finally that metals are not developing organisms but unchanging elements.

In this account of Helmont's developmental matter-theory, the word 'spirit' has appeared, and it could hardly have been avoided. The same word—or its forerunners, *spiritus* and *pneuma*—can be found in use from ancient times right down to 1700 and beyond. In their physiological writings, indeed, even such modern-minded men as René Descartes and William Harvey could not do without the term. Descartes attempted to produce a completely mechanistic physiology; but he retained the traditional 'vital, animal and natural spirits' in his account of the workings of the animal-frame; and neither he nor Harvey saw beyond the simplest anatomical picture to the serious beginnings of biochemistry.

The effective disappearance of the term 'spirits' from serious discussions of the nature of matter during the eighteenth century is a good index of the fundamental change in ideas which I am trying to make explicit here. For the reappearance of this term is a fair indication that the older, developmental paradigm of material change is still influential; and the crucial step in eighteenth-century matter-theory was the replacement of this developmental paradigm by a different one.

During the hundred years following Helmont and Newton men acquired a new familiarity with gaseous materials, and the idea of a gas soon lost all the spiritual overtones it had had for Helmont. One might say: in the eighteenth century the spirituous and the spiritual were finally distinguished. By the end of the eighteenth century all serious chemists accepted the fundamental conception of inert, unchanging chemical substances which we take for granted today. This was true both of the phlogistonians such as Priestley and of the oxygenists lead by Lavoisier. So much was this so, that the few remaining defenders of the older view regarded both Priestley and Lavoisier as equally misguided. Lamarck, for instance, criticized them equally: in his view, dynamic evolution had a part to play in physics and chemistry quite as fundamental as that it possessed later in his system of zoology. Dr. C. C. Gillispie has recently expressed Lamarck's attitude as follows:

> 'The business of physics (in Lamarck's view) is to bring each problem back to some inherent principle or tendency to perfection, carried out by the agency of a subtle fluid, which is distinguishable only in its effects. . . . The chemist must describe how active principles permeate and alter bodies in reaction.
>
> Lamarck's philosophy, therefore, is . . . a medley of dying echoes: a striving towards perfection; an organic principle of order over against brute nature; a life process as the organism digesting its environment.'

The doctrine of the fixity of species (Lamarck pointed out) was becoming a fundamental assumption of chemistry, just when serious evidence was appearing to overthrow it, at any rate in the field of zoology.

. . . . .

If we look back at the older style of explanation in matter-theory without trying to understand the fundamental model which served both Aristotle and Lamarck as the paradigm of material change, we shall certainly fail to sympathize with them. To twentieth-century eyes Aristotelian explanations have two great defects: either they appear entirely verbal—the *virtus dormitiva* once more—or else they look ridiculously purposive, as though Aristotle had attributed to all material objects a semi-conscious desire to turn into something different. To judge his explanations fairly, we need to see him as doing something far more legitimate; even though it was something we no longer consider fruitful. He is seeking to explain all material changes by 'placing' them at the appropriate position in some 'life-cycle'. The specific life-cycle for him is the explanatory unit—a seedling will turn into the corresponding adult plant, unless forcibly prevented, *because* it is (say) a cabbage seedling rather than a wallflower one. The natural thing to expect is that, if left to themselves in their usual environment, things will follow out their normal courses of development: if they fail to do so, the question will then arise how they came to be diverted from this typical sequence. And this question is a perfectly proper one: it even echoes Newton's own question, why the planets travel in closed orbits, and are 'drawn off from the rectilinear courses, which, left to themselves, they should have pursued'.

If we still judge Newton's theoretical question to be a

genuine one, while rejecting Aristotle's as unilluminating, this cannot be because the question is intrinsically misguided. Behind that judgment there lies, in fact, a wider one: for us, Newton's conception of inertia is a fruitful basis for the interpretation of all movement in Nature, whereas we no longer share Aristotle's conviction that all material changes are varieties of organic development.

We in 1960 can no longer afford to dismiss Aristotle's position too brusquely. For a century, the classical theories of Lavoisier and Dalton encouraged men to think of the distinction between 'animate' and 'inanimate' things as absolute. More recently, this distinction has come to appear less black-and-white. By now, indeed, many biochemists insist that the line between living organisms and inert chemical substances can be drawn only arbitrarily. Viruses and genes, they say, are only highly complex molecules. But an intellectual road cannot be opened to one-way traffic only. If the distinction between organisms and molecules is, after all, an arbitrary one, then the same must be true of the distinction between molecules and organisms. Any arguments which justify biochemists in speaking of genes as 'molecules of extreme complexity' justify us also in speaking of atoms and molecules as 'organisms of extreme simplicity'. And, in recent discussions about the origin of life on the Earth, Professor Calvin of the University of California has even spoken of 'natural selection' as having operated on the catalysts in the Earth's atmosphere, at the crucial stage when the first amino-acids were beginning to be formed.

Once again, what hampered Aristotle most in his matter-theory, as in dynamics, was not his being an airy philosopher: it was his very down-to-earthness. Things in the world around us are never composed of stuff which is, by chemical standards, 100 per cent pure, sterile, and inert. For the earliest thinkers, water was the same as rain, and this was not inactive H_2O: it was the spermatozoa of the Sky-god, by which the Earth-goddess was caused to conceive and bring forth crops. Even today, producing chemical substances of extremely high purity is still a difficult matter; and, once they are produced, chemists keep them carefully in bottles fitted with ground-glass stoppers to prevent contamination.

Why is this? It is because our idea of an absolutely pure, inanimate, unchanging material stuff is scarcely to be realized, except by artificial means: examples are almost as rare as examples of absolutely pure inertial motion. This was the reason why I declared, at the beginning of the present chapter, that the fundamental concepts of our modern matter-theory were *idealized*, in the same way as the fundamental concepts of Newton's dynamics.

One last point may be made, to draw together the threads of this discussion. Our 'ideals of natural order' mark off for us those happenings in the world around us which do require explanation, by contrasting them with 'the natural course of events'—i.e. those events which do not. Our definition of the 'natural course of events' is therefore given in negative terms: positive complications produce positive effects, and are invoked to account for deviations from the natural ideal, rather than conformity to it. This being so, the appearance of words like 'inert' and 'inertia' in our theories is perhaps significant. For these are essentially negative terms, indicating how things will behave of themselves, if nothing is done to them from outside.

The question is: how are we to tell when 'nothing' is being done? The answer to this question too is at the mercy of history, and changes with everything else at the fundamental level of theory. Suppose the traces by which a horse is pulling a cart happen to break, so that the cart comes to a stop. This sudden halt will be interpreted by Aristotle as entirely 'natural', being a consequence of the fact that the horse's strength is no longer being exerted on the cart. Of course the cart will stop; nothing is acting on it, so its 'inertia' immediately brings it back to rest. Galileo and Newton, however, have taught us to see the relation between the horse and cart differently. Left to itself, the cart would keep rolling along the level for ever: the horse is simply a necessary device for overcoming the effects of friction, gravity, and air-resistance. The fact that the cart comes to a stop after the traces break is thus, to our modern eyes, a *positive* phenomenon: if *nothing* were acting on it, it would keep moving, so that friction or something must be stopping it.

So again in matter-theory: if we regard the normal development of things through their life-cycles as entirely self-explanatory, we may end up with a science of pathology, but we shall place less importance on biochemistry. For, in that case, all we shall require to have explained will be why the development of things departs from its natural course. We shall take the fixed life-cycles of different creatures for granted, in the way nineteenth-century chemists took the fixed nature of the chemical elements for granted. And perhaps the emancipation of matter-theory from the older system of ideas could be complete only when men recognized the proper relation between physiology and pathology—i.e. since the work of Claude Bernard. From the pathological point of view, the proliferation of cancer-cells is something 'unnatural', being a deviation from the standard physiological function. For the biochemist, however, the task is to identify all the processes going on in the body equally, functional and dysfunctional alike. What we call 'normal' physiological processes are a special case, which to the biochemist are neither more nor less 'natural' than pathological ones. So, at the biochemical level, the only theory giving complete understanding will be one which treats normal and pathological processes on a basis of complete equality.

. . . . .

To conclude: in studying the development of scientific ideas, we must always look out for the ideals and paradigms men rely on to make Nature intelligible. Science progresses, not by recognizing the truth of new observations alone, but by making sense of them. To this task of interpretation we bring principles of regularity, conceptions of natural order, paradigms, ideals, or what-you-will: intellectual patterns which define the range of things we can accept (in Copernicus' phrase) as 'sufficiently absolute and pleasing to the mind'. An explanation, to be acceptable, must demonstrate that the happenings under investigation are special cases or complex combinations of our fundamental intelligible types.

Some of our explanatory ideals claim universal application: this is true both of 'inertial motion' and of 'chemical substance'. It sometimes happens, however, that a particular ideal has a more restricted application. A pattern-of-theory or form-of-explanation may be, not uniquely right, but appropriate to one range of studies rather than another. We shall consider some examples in the next chapter. As we shall see, there may be no way of telling beforehand in what field a particular pattern of explanation will bear fruit. If this is borne in mind, many of the seemingly pointless investigations of past scientists acquire a new meaning. One cannot even label a false trail as such without exploring it some way first. There is nothing unscientific about these explorations: it would be unscientific to neglect them. The scientific ideas which survive are the ones which have best proved their worth, while those which have been discarded—for example, the ideas of the alchemists—can be thought of as the pterodactyls of science.

5

Forms and Styles of Theory

THE German philosopher Immanuel Kant had a life-long interest in astronomy and physical theory. He began his career by writing about physics rather than metaphysics, and even in his later work his earlier interests shine through. As an introduction to his *Critique of Pure Reason*, for instance, he wrote the shorter and easier *Prolegomena to any Future Metaphysics*: in this essay, at the end of a section entitled 'How is Pure Science of Nature Possible?', he discusses the phenomenon of gravitation:

> 'If we proceed . . . to the fundamental teachings of physical astronomy, we find a physical law of reciprocal attraction applicable to all material nature, the rule of which is that it decreases inversely as the square of the distance from each attracting point—that is, as the spherical surfaces increase over which this force spreads—which law seems to be necessarily inherent in the very nature of things, and hence is usually propounded as knowable *a priori*. Simple as the sources of this law are, merely resting upon the relation of spherical surfaces of different radii, its consequences are so valuable with regard to the variety and simplicity of their agreement, that not only are all possible orbits of the celestial bodies conic sections, but such a relation of these orbits to one another results that

> no other law of attraction than that of the inverse-square of the distance can be imagined as fit for a cosmical system.'

This passage provides both an elegant scientific illustration and a clouded philosophical vision of the part played in scientific understanding by intuitively intelligible forms of theory. We can use it as a text for the present chapter, whose topics are the questions: How far is the adequacy of a scientific theory determined by its formal pattern? And how do we know what patterns of thought are appropriate in any particular field of study?

I shall argue for the following conclusions. Over particular forms of theory, black-and-white questions of truth and falsity do not arise. The crucial issue is, rather, what types of happening a particular form of theory will help us to understand and where its use will be unhelpful or misleading. A form of theory which has been applied successfully in one field is sometimes introduced into another with equally fruitful results; yet at other times things work out differently. A form of theory previously fruitful in one field may lead men down a blind alley in another, or alternatively, a form of explanation which had earlier obstructed understanding in one field may prove extremely fruitful in another. Nor need there be anything to indicate beforehand which way things are going to turn out.

. . . . .

Let us try to state Kant's problem exactly. The point at issue is this: Newton had established in the *Principia* that his inverse-square law of gravitational attraction would explain a multitude of natural phenomena, and had concluded that the law in fact applied universally. Every massive body throughout the whole universe (he inferred) is continually acted upon by forces directed towards every other such body, the force on any unit mass increasing with the mass of the attracting body and falling off as the square of its distance. Newton produced a great deal of observational evidence to support this supposition, but beyond that point he was unable to go. He could demonstrate no mechanism to account for the gravitational interaction; and he could offer no further reason why the attraction should vary as the inverse-square, rather than (say) the inverse-cube of the distance. As he left the matter, the inverse-square equation was something whose applicability to nature had to be accepted as a brute fact.

This, in Kant's view, was not satisfactory. Surely, he thought, there must be some further reason why this simple and elegant law applied. Its very elegance and simplicity demanded further explanation.

Kant thought he had found a deeper explanation, and gave it in the passage quoted. 'The rule [he says] is that the reciprocal attraction decreases inversely as the square of the distance from each attracting point', and this amounts to saying that the force drops off 'as the spherical surfaces increase over which this force spreads'. For, of course, two spheres with a common centre but different sizes will have surface areas proportional to the squares of their radii: if the outer one has twice the linear dimensions of the inner one, for example, its surface-area will be four times as great. Here was the clue for which Kant had been looking. The further you go from the attracting-point, the more thinly the gravitational action spreads out. The increasing areas of the spherical surfaces must be counter-balanced by a weakening of the force: so, he concluded, 'the sources of this law rest clearly upon the relation of spherical surfaces of different radii'. This discovery convinced him that 'no other law of attraction can be imagined as fit for a cosmical system'.

Evidently, Kant saw something intrinsically intelligible in the inverse-square law, and thought that he had captured it in this argument. Yet nowadays few physicists would follow him in this. His argument assumes too much. True, it has its attractions. If we could believe that gravitational force spread out from the sources of attraction like an indestructible fluid, then—inevitably—the further out one went, the thinner it would be spread over the surfaces it crossed. There would be (so to speak) only so much force to go round, and the strength of the force would indeed depend on 'the relation of spherical surfaces of different radii'. But the validity of this argument depends on the assumption that gravitational force is propagated and conserved, in the way in which matter and energy can be. Kant neither gives a further justification for this assumption, nor even explicitly states it.

We may at first be tempted to dismiss this argument as a mere blunder. For, on general grounds, we might have expected some other quantity—for instance, force times distance—to be governed by a conservation law, rather than force itself. We may concede that it was a pardonable blunder, since Kant was writing seventy years before a satisfactory distinction between force and energy had been formulated: nevertheless, something about the argument strikes us as quaint.

Mistakes, however, can sometimes be more interesting than successes. For where exactly does Kant's mistake lie? Certainly there is nothing wrong with the *form* of his argument. Formally, it is identical with other arguments which we learn to this day in elementary physics courses. For example, a very early proof in the elementary theory of light leads to an equation for the 'intensity of illumination' produced on a surface at a distance from a point-source of light: in the absence of any absorbing medium, this intensity (it states) will drop off inversely as the square of the distance from the point-source to the illuminated surface. The proof consists in constructing two supposed spheres of different sizes around the point-source, and arguing that light propagated from the source in a given direction will cross areas of each sphere increasing as the squares of their radii. By the time it has gone twice the distance, any given quantity of light will be spread out one-quarter as thickly, and so on for other distances. *Ergo*, the intensity of illumination *must* vary inversely as the square of the distance from the point-source.

So far as its form goes, this argument is identical with Kant's argument about gravitation. Formally speaking, therefore, the one argument is as good or as bad as the other. In each case we are presented with the same fundamental model, of an 'influence' spreading uniformly out from a point and becoming more thinly spread as the distance increases. The rest follows. Why, then—if there is nothing formally to choose between the proofs—does Kant's argument strike us as quaint, and the other appear almost axiomatically correct?

The answer to this question becomes clear only if we look at the theoretical backgrounds of the two arguments. There was no reason, in Kant's time, to think that this form of argument must in principle be applicable to light and not to gravitational force. Consider the situation in optics. Nowadays, we begin the study of light by defining intensity of illumination and proving *a priori* that it falls off as the inverse-square, rather as mediaeval scholars defined uniform acceleration and proved *a priori* that it obeyed the rule 'Distance increases as time-squared'. In this way we can build up a natural and rational way of thinking about illumination, and define a number of useful technical terms. The value of this system of thought is subsequently reinforced by two other things: first, our success in devising practical methods for measuring illumination, as a result of which we can apply our initial definitions and theorems; and secondly, the deeper foundation given to the subject by Maxwell's theory of electromagnetism, which interpreted light as a form of electromagnetic energy, and so justified the assumption of its conservation.

In gravitation theory matters went very differently. Kant's proof acquired neither a pragmatic value nor a theoretical basis. Gravitational forces have always been notoriously difficult to measure, and we can scarcely manipulate gravitation to serve our purposes, in the way we do illumination. Intellectually, too, the whole character of gravitational force is still almost as problematic now as it was when Kant wrote. Had the history of physics since 1800 gone differently, his argument might have been given the same permanent and honoured a place in the elementary textbooks as its counterpart in illumination-theory. Unfortunately for Kant, the steps by which the latter argument established itself in optics have so far not been taken in the case of gravitation.

The situation might yet change. In the last year or two, in fact, there have been a few significant straws blowing in that direction. For instance, the great Cambridge physicist P. A. M. Dirac has revived speculation about the 'quantization' of the gravitational field, suggesting that we should accept the idea of 'gravitons' (corpuscular packets of gravitational energy) corresponding to the 'photons' of electromagnetic theory. Suppose that these new suggestions were now to receive striking new observational support. Our attitude towards Kant's proof might change suddenly. Instead of being condemned as 'a silly blunder', it would be then in danger of being praised as 'a brilliant anticipation'. In actual fact, I am arguing, it was neither the one nor the other.

. . . . .

Let us now look at an example of the reverse process: a case in which arguments of a form discredited in one field of science (chemistry, to be precise) bore unforeseen fruit later in a different one—namely, genetics.

During the century before Lavoisier, writers on chemistry made much use of a conception which we have since abandoned, which they frequently expressed by the use of the word 'principle'. Throughout the writings of the period one comes across recurrent allusions to the 'principle of inflammability', the 'metallizing principle', and the like: sometimes it seems as though there were as many 'principles' as phenomena—or 'spirits' earlier. Curiously enough, historians of chemistry generally pass over this keyword in silence, leaving us to discover for ourselves what its use implies. Yet this is a pity, since the assumptions lying behind this use are interesting and significant enough to be worth bringing into the open; and, in recognizing why these ideas were abandoned in subsequent work, we come to understand better the novel features of classical nineteenth-century chemistry.

I will first try to state the general problem with which eighteenth-century chemists were concerned, and then indicate how the notion of a 'material principle' arises naturally out of this problem. Chemical substances, like living creatures, can be arranged in a taxonomic system. To begin with, we class them together in large groups—as acids, salts, bases, metals, and so on. These groups are identifiable by easily recognizable properties which their members share. Metals have a sheen, are ductile, and conduct heat well; salts tend to be crystalline, acids acid, and so on. These large classes of substances can be compared to the genera and families of botany and zoology. Within each genus, there are distinct species: the class of metals comprises gold, silver, iron, lead, mercury, copper, etc.; the class of acids contains vitriolic, muriatic, nitric, carbonic, and other acids; and likewise for the other groups. Each specific acid or metal then possesses some properties common to all acids or metals, and these may be called its 'generic characters'; while other properties —its 'specific characters'—mark it off as gold rather than silver, or vitriolic rather than muriatic acid. All this was clearly known by the early eighteenth century. Once this was recognized the question naturally arose: what connection is there between the macroscopic qualities which determine the taxonomy of chemistry (so to speak) and the material constitutions of the different substances? What ingredient, or mode of combination of ingredients, determines that a particular piece of matter shall be (say) metal by genus and gold by species?

That was the fundamental question about material substances for eighteenth-century natural philosophers—for Lavoisier as much as his predecessors. Lavoisier was particularly interested in the acids, and had a very definite idea about how they were composed. Every acid consisted, in his opinion, of two elementary substances: one of them 'acidifying', the other 'acidifiable'. The first of these elements was present in the make-up of all acids, and for this 'acidifying principle' or acid-begetter he coined the name *oxy-gen*; its special role was to confer on all acids their generic properties. The specific properties of any particular acid, which marked it off as sulphuric (vitriolic) rather than carbonic or nitric, came from the other ingredient in the compound, which might be any of the possible acidifiable substances—e.g. sulphur, carbon, or azote (nitrogen). To state his theory briefly: the generic characters of an acid are determined by the acidifying principle, the specific characters are conferred by the acidifiable principle. Each character or set of characters is related to, and can be explained in terms of, a corresponding ingredient.

This pattern of thought is a familiar and natural one, which frequently reappears, and is still associated with the word 'principle'. It is appropriate wherever some recognizable property or group of properties can be conferred on a body, or can be removed from a body, by adding or extracting a definite ingredient. It has its place in the kitchen and the bar: salt makes things salt, bitters make them bitter—there, in fact, the properties and their conferring 'principles' can even share names. Even in more technical circles, this idea still has a place. A pharmaceutical manufacturer will often advertise a drug in the medical journals as containing (for instance) digitalin, 'the active principle' of the foxglove, or as having therapeutic properties of a kind which justifies calling it (say) 'an anti-pyretic principle'. *Quot ingredientia, tot qualitates*: to get the quality, add a pinch of the principle.

If we move on from eighteenth-century chemistry—even Lavoisier's theory—to the classical chemistry of the nineteenth century, there is a striking change: this term 'principle' disappears from fundamental discussions. Once Dalton's atomic theory was developed and applied systematically to explain the facts of chemical change, one thing became crystal-clear. The overt, easily-recognized properties of chemical substances did not go—one for one—with particular material ingredients: the manner in which the elementary ingredients were combined was, in most cases, much more important. Sulphur conferred no single property or group of properties on all sulphides, sulphates, and other sulphuric and sulphurous compounds alike—unless it was the fact that, by suitable treatment, one could extract sulphur from them. Lavoisier's own pet 'acidifying principle'—namely, oxygen—was not even present in all acids.

Muriatic acid, it turned out, contained only hydrogen and chlorine, and was eventually rechristened hydrochloric acid. The overt properties of chemical substances were evidently related to their material composition in a more complicated way than eighteenth-century chemists had supposed. So it became necessary, in all research from then on, to establish the material constitution of a compound in the first place, quantitatively and independently of its manifest properties. Only when this chemical recipe was known could one ask what correlations there were between manifest macroscopic characters on the one hand and inferred elementary constituents on the other.

The programme of nineteenth-century chemistry proceeded on this new basis; and it broke entirely, at the theoretical level, with the idea of 'material principles'. Yet this was not the end of the story. The mode of argument was not banished from science for ever. The idea behind the term 'principles'—that macroscopic properties or characters are to be related directly to corresponding microscopic constituents—was not entirely swept out of scientific theory into the limbo of the pharmaceutical advertisements. Not at all: this idea, which served eighteenth-century chemists first as a crutch, but later as a broken reed, came to life again in Mendelian genetics.

The mode of thought we are studying was, in fact, fundamental to Mendel's theory. His predecessors had thought of a species or variety of plant as an organic unity, whose nature was transferred from parent to offspring indivisibly. Working along those lines, they had thrown little serious light on the mechanism of inheritance. Mendel saw an alternative approach. He regarded each plant as a 'mosaic' of characters rather than a unity, and associated each overt character with some corresponding internal 'factor'; then, appealing to the results of his famous experiments in hybridization, he demonstrated that —in many cases at least—different characters are 'segregated' and transmitted independently in statistically predictable proportions; and these statistical discoveries lent themselves at once to explanation as resulting from the transmission of corresponding microscopic 'factors'.

Mendel, it is true, does not actually apply the word 'principle' to these factors, yet the form of conception he is using is precisely that we met with in earlier chemistry. For eighteenth-century chemists, a particular substance was a 'mosaic' of properties, such as acidity, conferred by corresponding 'principles'; so now, for Mendel, a particular living creature was an assemblage of characters associated with, and transmitted by, corresponding ingredient factors. In due course Mendel's factors became Johanssen's 'genes'. It has even become the practice in genetics to *name* particular genes, microscopic though they may be, according to the macroscopic properties they are supposed to transmit: e.g. to speak of the gene 'red-eye', or the gene 'short-wing', as we speak of 'salt' and 'bitters'. The whole brilliant theory of genetics created by T. H. Morgan and his school can, in this way, be thought of as a highly successful application to inheritance of the intellectual pattern which in chemistry Dalton's work had discredited.

In genetics, as in chemistry, this intellectual situation was stable only so long as it was hard to study the hypothetical ingredients directly. But history now looks like repeating itself further. Once the Mendelian factors had been associated with particular material parts of the cell-nucleus, cytological genetics could join hands with statistical population-genetics; and, for some years, the alliance was an amicable one. Inheritance proved frequently to take place on the 'principle' pattern to a surprisingly high degree of accuracy. Yet the programme of treating the gene in a double way, as 'a unit of function and a unit of structure as well', was bound eventually to lead genetics on from its Lavoisierian into its Daltonian phase. In the last few years, in fact, geneticists have been feeling their way beyond the classical theory. The fundamental procedure of correlating overt characters directly with microscopic constituents, already modified by the idea of 'buffering', has lost its theoretical status. Once begun, these changes will probably continue. The more that nineteenth-century chemists discovered about the chemical elements by direct study, the more complex became their ideas about the relationship between properties and ingredients; and, presumably, the better we understand the biochemical structure of the cell-nucleus itself, the less exact will Mendel's original assumptions about inheritance prove. These assumptions have probably borne more scientific fruit than their eighteenth-century counterparts in chemical theory; yet perhaps they have had their day.

. . . . .

In these studies of the nature of explanation I have not set out to analyse my chosen scientific episodes in great historical detail. I have aimed only at showing the fascinating problems that arise when one brings logical and philosophical questions to bear on the history of our scientific ideas. The intellectual frame of a man's thought displays itself less in the detailed results he enunciates than in the questions he asks and the assumptions which underlie his theorizing. So it is pointless—and boring—to go through past science chronologically, awarding high or low marks to our predecessors simply by asking whether they assert or deny doctrines similar to those we still accept today. Against different intellectual backgrounds, one and the same ostensible doctrine may take on very different complexions.

For example: Robert Boyle is often congratulated on adopting the corpuscular philosophy, the view that there are different material elements, each with its own kind of atom. Yet, by itself, Boyle's atomism did him little good. For he was still prepared to think of fire as a substance, which could become 'fixed' in bodies and contribute to their weight; and this belief stood in the way of his making crucial discoveries and distinctions. The atomism of John Dalton was fruitful for chemistry, as Boyle's corpuscular theory never was, just because Dalton could build on the previous analysis of Lavoisier, from which it was clear that fire could not be made (in Boyle's words) 'stable and ponderable'. In Newton, Boyle, and even Lavoisier—to say nothing of Democritos—atomism was part of a general philosophical position: only in Dalton does it really become a chemical theory.

It is simply irrelevant to criticize earlier scientists for blindness, when the doctrines they failed to accept had no intelligible place within the theoretical framework of their time. If they sometimes seem to us now to have shut their eyes to the facts, we need to ask: 'Could these things which we regard as "facts" have been certain, clear, or even intelligible to earlier investigators?' The continual interaction of theory with fact—the way in which theories are built on facts, while at the same time giving significance to them and even determining what are 'facts' for us at all: that, for the philosophically minded student, is the extraordinary fascination and delight to be had from reflecting on the historical evolution of scientific ideas.

In this chapter our subject has been *forms* of scientific theory, but let us also look briefly at the associated question of *style*. At this point I want to raise a question rather than answer it. Suppose, then, that we investigated the development of ideas about magnetism over the last five centuries. (Such a study could be very valuable, and needs undertaking.) In the course of this enquiry a problem would have to be faced: let me first paint in the background.

About the basic facts of magnetism, only a certain amount can be said. Any magnetic body, from a lodestone to an atomic nucleus, exerts forces on other magnetic bodies at a distance whose effect is to attract or repel or rotate them according to a familiar pattern, long understood. Up to a certain point at any rate, the physics of magnetism, too, is by now pretty well understood: we can explain how sub-microscopic magnets combine together to give a material object its macroscopic magnetic properties, and the manner in which electrical and magnetic effects act and react on one another is also well established. Yet about the unit-magnets themselves we are still very much where Gilbert was in 1600 when he studied large-scale magnetic bodies. We can say what effects they have on other magnetic bodies, how these effects vary with distance and orientation, and how they depend on the magnetism of the earth itself; but that is about all. Our fundamental beliefs about magnetism have, in fact, changed very little in the last three-and-a-half centuries, even though the linguistic idioms which physicists have used in describing magnetic phenomena have changed frequently. At one time, these phenomena were put down to the action of 'magnetic effluvia'. At a later time, 'subtile fluids' of a special magnetic character were credited with the same effects. These were succeeded later by magnetic corpuscles, magnetic fields, and even by bare, unattached mathematical functions.

Now, a historian of magnetism will have to ask himself the following logical question: 'How far have these changes in idiom been responses to brute experience, reflecting justifiable inferences from experience or observation? How far have they reflected the changing theoretical affiliations of magnetism? Or have these idioms sometimes altered rather for aesthetic reasons, as matters of intellectual fashion, along with the wider modes in which men's creative interpretation of Nature has found expression?' I suspect myself that only a small proportion of changes in magnetic theory have had any direct empirical justification, and followed on the discovery of magnetic phenomena explicable in terms of (say) 'subtile fluids' but not 'effluvia'. More have sprung from the second cause—following, for instance, on the integration of magnetic theory with electrical and optical theory by Clerk Maxwell. Yet, even when both these kinds of changes are allowed for, a residuum will probably remain. If pursued beyond a certain point, the study of scientific ideas may thus bring one up against questions of style, or even of mere fashion. The question is at any rate worth investigating.

A single experiment can answer a single question. But this question has to be framed, and it must be framed in terms of a provisional theory. If enough of the questions to which a given theory leads us turn out unprofitable, we of course seek to abandon it in favour of a better. But it is not easy to expose the fundamental intellectual frameworks of an age to the bare mercy of the facts. Our tentative hypotheses and theories are the hybrid offspring of the things we have studied and the general ideas which have regulated our interpretation of their behaviour. At the fundamental level, where the question at issue is the style of interpretation favoured in a given century, historians of scientific ideas still have much to bring to light.

6

The Evolution of Scientific Ideas

SCIENCE is not an intellectual computing-machine: it is a slice of life. We set out on our enquiry into the aims of science, hoping to do two things: first, to define in a life-like way the common intellectual tasks on which scientists are engaged, and the types of explanation their theories are intended to provide; and secondly, to pose the problem, how we are to tell good theories from bad, and better ideas, hypotheses, or explanations from worse ones.

We began by scrutinizing one popular answer to these questions: the 'predictivist' account. It soon became clear that this account would not do all we wanted. Scientists are concerned with 'forecasting-techniques' only incidentally; and any more satisfactory sense of 'prediction' takes for granted the idea of explanation, rather than defining it. The central aims of science are, rather, concerned with a search for understanding—a desire to make the course of Nature not just predictable but intelligible—and this has meant looking for rational patterns of connections in terms of which we can make sense of the flux of events. So we have placed in the centre of our enquiry two questions: 'What patterns of thought and reasoning give scientific understanding?', and: 'What factors determine which of two rival theories or explanations yields greater understanding?'

The first question brought us up against the basic conceptions or 'ideals of natural order', which settle what a scientist regards as 'self-explanatory' or 'natural'. At any stage in the evolution of science, I argued, certain forms of explanation present themselves to men as being entirely intelligible—e.g. the standard, though different, types of motion accepted as self-explanatory by Aristotle, Galileo, and Newton, and also such ideas as that of a 'pure substance' or a 'typical life-cycle'. The scientist may start with half a dozen different ideal structures or processes, yet in one way or another he must put them to work if he is to make the course of Nature intelligible.

For the logician, these explanatory ideals pose a particular problem. On the one hand, they change and develop, as time goes on, in the light of discovery and experience: so they must be classed as 'empirical', in a broad enough sense of the term. On the other hand, one cannot confront them directly with the results of observation and experiment. They have to prove their worth over a longer term, in a way which still needs analysing. Though changing with history, they are also —for the individual scientist—'preconceived' notions: thought out beforehand, and applied only subsequently to particular scientific problems.

Yet, if one speaks of these conceptions as preconceived, a distinction must be made, for they are 'preconceived' in a perfectly innocent sense of the word. Scientists are rightly suspicious of 'preconceived ideas' and pride themselves on coming to Nature in a spirit of objectivity. If a man enters the observatory or laboratory with preconceived ideas about what he will find, this (scientists feel) will prejudice his investigation. If he has already made up his mind (say) that pigs *can* fly, that will disqualify him as an observer: he will go around the world looking for evidence to support his prior belief, and may end by hailing some porker leaping off the roof of his sty as proof of his contention. So far, the scientists are undoubtedly in the right: when it comes to interrogating Nature, in the laboratory or in the field, we must leave her to answer for herself—and answer without any prompting.

That, however, is not the point at which our 'ideals of natural order' come in. Their influence is felt earlier. For, though Nature must of course be left to answer to our interrogations for herself, it is always *we* who frame the questions. And the questions we ask inevitably depend on prior theoretical considerations. We are here concerned, not with prejudiced belief, but rather with preformed concepts; and, to understand the logic of science, we must recognize that 'preconceptions' of this kind are both inevitable and proper—if suitably tentative and subject to reshaping in the light of our experience. If we fail to recognize the conceptions for what they are, we shall not appreciate the true character of our scientific ideas, nor the intellectual problems which faced our predecessors, through whose labours our own ideas were gradually formed.

. . . . .

There is only one way of seeing one's own spectacles clearly: that is, to take them off. It is impossible to focus both on them and through them at the same time. A similar difficulty attaches to the fundamental concepts of science. We see the world through them to such an extent that we forget what it would look like without them: our very commitment to them tends to blind us to other possibilities. Yet a proper sense of the growth and development of our ideas will come only if we are prepared to unthink them. We are *justified* in placing the trust in them that we do, only because—and to the extent that—they have proved their worth in competition with alternatives: if earlier men had never thought in other terms than we do, then we ourselves would simply be carrying on a traditional habit. We shall understand the merits of our own ideas, instead of taking them for granted, only if we are prepared to look at these alternatives on their own terms and recognize why they failed.

The story we looked at in Chapter 4 will serve as an illustration. In many histories of science, the account which is given of the development of chemistry before 1650 stands in sharp contrast to the account given of earlier cosmology and dynamics. By now, the paths which lead down, from Babylonia and Ionia through Athens, Alexandria, and Baghdad to Copernicus and his successors, are well charted and understood; and, as a result, historians can tell a coherent and reasonable story about this strand of intellectual history. Turning to matter-theory, we find not order but chaos: we get an impression of ideas developing haphazard, leading through a sequence of deplorable errors and byways into a morass, from which the seventeenth-century corpuscular philosophers alone escaped.

But was this picture of earlier chaos a fair one, after all? The historians in question were often deeply committed to the inorganic chemist's point of view, and found it easy to focus only on a restricted range of ideas. We can now see, however, that matter-theorists before Newton and Boyle were often operating on quite a different intellectual plane: it was always vain trying to get their speculations into focus without laying aside our atomic presuppositions. Material change was, for many of those earlier thinkers, primarily a physiological matter. Only in recent centuries have men generally reversed the direction of thought, and explained physiological processes themselves in terms of chemical ideas. The possibility—not to say the necessity—of doing this took a good deal of establishing. Yet we are now so accustomed to the new picture that we find it hard to think back and see the world of natural change through the eyes of earlier men—for whom the natural object of comparison was the organism, with its characteristic cycle of birth, adolescence, maturity, and decay. This is certainly one reason why the history of matter-theory before 1650 appears in retrospect so chaotic and haphazard.

To a lesser extent, the same thing has happened in the history of dynamics. We contrasted in Chapters 2 and 3 the different ideals of natural motion to be found in Aristotle and Copernicus, Galileo and Newton. Copernicus was in almost every respect still an ancient scientist—in particular, an Aristotelian one. The natural motion of the heavenly bodies was for him, as for Aristotle, uniform circular motion. By contrast, Newton employs throughout his work a quite different paradigm of natural motion: it is, to be sure, still motion-at-uniform-speed, but it takes place in straight lines rather than in circles. This new conception of inertial motion had in his time been under discussion for less than a century; yet already one finds him taking it for granted. He seems not to have recognized that, in this respect, he was departing radically from all ancient views. Rather, he implies that the ancients themselves must have thought in the same way:

> '*We do not know* in what manner the ancients explained the question, how the planets came to be retained within certain bounds in these free spaces, and to be drawn off from the rectilinear courses, which, left to themselves, they should have pursued, into regular revolutions in

> curvilinear orbits. . . . For, from the laws of motion, it is most certain that these effects must proceed from the action of some force or other.'

In a sophisticated man one might take this for pretended ignorance, assumed in the interest of elegant exposition. In the case of Newton that supposition would be out of character; and his words are, rather, evidence of the speed with which men can accept a radically new vision of the world as 'second nature'.

The invisibility of our intellectual spectacles may have a further effect. It may lead us to misunderstand, not only the specific ideas and doctrines put forward by earlier thinkers, but also the general character of their enquiries. This effect, too, is easily illustrated: for example, from the scornful third-hand accounts one reads of intellectual history in the days before the Scientific Revolution. 'What were thinking men up to in all those years? And why did they shut their eyes to the merits of the experimental method? How could they have hoped to get genuine scientific results from mere argument, without leaving their studies, unless it was through a mixture of prejudice, muddle-headedness and metaphysics?' Such questions as these often criticize the 'pre-scientists' for *failing* to do things that it was not their business to attempt. For (as I want to argue next) those earlier enquiries in natural philosophy that are swept aside as 'pre-scientific' were in fact indispensable. Their effect was to clear the ground, and collect many of the girders and timbers out of which the structure of science as we know it was in due course constructed.

Two examples will indicate how easy it is to praise or blame earlier thinkers for irrelevant reasons. Consider, first, the Greek natural philosophers: those men in the cities of

Asia Minor, and later in Athens, who began to theorize about Nature in something recognizably like our own scientific way. In thinking about their ideas, we are subject to two opposite temptations. We may be so struck at finding in their teachings patterns of thought which bore scientific fruit many centuries later that we hail them as 'brilliant anticipators'. Democritos the atomist is frequently the subject of this kind of praise. During the hundred years between Dalton and Heisenberg the atomic theory was regarded by many as the most profound and fundamental truth about Nature; so that any man found proclaiming it more than two thousand years before Dalton seemed to be a scientific genius. Since the development of wave-mechanics, the situation is somewhat changed. Classical atomism is dead, and theoretical physicists can discuss quite seriously whether so-called 'fundamental particles' might not be replaced by mathematical singularities in fields of force—a conception having more in common with the continuum theories of the Stoics than with the unvarnished atomism of Democritos.

On the other hand, when looking at the methods of investigation used by natural philosophers in ancient Greece, rather than at their results, we may be equally tempted to over-scorn them. Taking up a modern methodological point of view, we may find their generalizations reckless, their understanding of experimental technique negligible, their arguments trifling. Why, for instance, did the Greek philosophers take so little trouble to put their theories on a sound observational basis? (I have in mind the pre-Socratics, or what we know of them; for it is scarcely possible to speak so harshly of Aristotle.)

Yet this criticism, too, misses the main point. Before one can put forward, and choose between fully-developed theories, one must first explore the possibilities opened up by different conceivable *types* of theory. It might, after all, be possible to rule out certain 'designs for theories' without going to the actual labour of constructing them: a design must show promise before one gets down to the details of building and testing. Many of the Greek philosophers were undoubtedly arguing at this preliminary stage; and the things they have to say on this subject are often extremely acute. Furthermore, they foresaw between them most of the general types of explanation that have served science well during subsequent centuries. To put the point briefly: what the Greeks established was not the actual nature of things, but the possibility of giving a rational account of Nature. They showed clearly, too, the extreme possibilities and limitations of various different approaches: so much so, that their discussions of fundamental matter-theory can still offer enlightenment to a working physicist like Werner Heisenberg.

Mediaeval natural philosophers, also, have long deserved to be rescued from irrelevant criticisms. They have often been brushed aside by polemical historians of science as 'tedious logic-choppers', who perpetuated Aristotle's errors as a result of failing to study the world at first hand. In recent years less prejudiced scholars have uncovered a very different picture. The scholastic philosophers may have been logic-choppers, but they chopped to good purpose. Where Aristotle left only a general theory of change, which provided no more than a bare foundation for mechanics, fourteenth-century mathematicians worked out for the first time a whole battery of important distinctions: e.g. between linear and angular, average and instantaneous velocity.

They advanced the first satisfactory definition of uniform acceleration, and proved the most important theorems which follow from this definition—including the one which was crucial for Galileo: that a uniformly accelerating body travels distances from rest proportional to the square of the time. They even developed the principles of measurement and calculation to the point of recognizing what form a satisfactory scale of temperature must take. Suppose one represents degrees of warmth along a line from an arbitrary point, then (they tell us):

> 'Let there be given water of two weights hot in the 6th degree; let there be given again another water of one weight hot in the 12th degree with respect to the same point; a mixture of the two waters having been made, the hotness of the mixture will be eight degrees, with respect to the aforesaid point, since the distance that is between six and eight is one-half the distance that is between eight and twelve, just as the water of one weight is half the water of two weights.'

Yet these doctrines were put forward purely in the abstract, as exercises in theory: the rules obeyed by a uniformly accelerating body were worked out mathematically, not demonstrated in experience, and the degree of warmth of a mixture was estimated by men living two centuries before the construction of a satisfactory thermometer.

Was their work, then, scientifically pointless? Did they, through their neglect of experiments, achieve nothing? If that is our conclusion, we are failing to recognize the debt we ourselves owe to these logic-choppers. For they built up a tradition to which Galileo himself was demonstrably an heir. The kinematical theory in his *Two New Sciences* bears a relation to their discoveries like that which Euclid's geometry bears to those of earlier Greek geometers. The mediaeval theorems were still being taught in Italy when Galileo was a student, and he continued to write in terms which he had inherited from them. Furthermore, when he described his famous experiments—rolling balls down inclined planes, and so on—the effect of his work was not to overthrow the mechanical theories of the mediaevals; rather he succeeded, in the spirit of a physicist, in showing the direct relevance to the world of Nature of theorems which they had worked out previously in the abstract. They were the natural philosophers, who built up the stock of ideas with which he worked: he was the physicist, who put these ideas to work in his explanations and explored their range of application.

If the growth of science has had a number of phases, demanding quite different methods of work, that should not surprise us. For the business of science involves more than the mere assembly of facts: it demands also intellectual architecture and construction. Before the actual building comes the collection of materials; before that, the detailed work at the drawing-board; before that, the conception of a design; and, before that even, there comes the bare recognition of possibilities. No wonder science has included, and must include, much *a priori* study of possible forms of theory, developed without immediate regard to the particular facts of Nature.

Unless these possible forms of theory are eventually applied to explain the actual course of events, our *a priori* studies will of course bear no positive scientific fruit. Yet they are a part, and a legitimate part, of scientific enquiry now as in previous centuries. What sorts of theory are admissible at all; what conceptions could possibly explain, in some field of study, the familiar facts of common experience; what bearing theories in different fields have on each other . . . the men who discuss speculative questions of these kinds play an essential part in the development of science, quite aside from the devoted labours of the white-coated laboratory workers. Indeed, the long-term rewards of successful speculation are greater than those of experiment. The greatest fame is reserved for those who conceive new frameworks of fundamental ideas, and so integrate apparently disconnected branches of science. Isaac Newton, Clerk Maxwell, and Charles Darwin are best remembered, not as great experimenters or observers, but as critical and imaginative creators of new intellectual systems.

. . . . .

We need, accordingly, to see scientific thought and practice as a developing body of ideas and techniques. These ideas and methods, and even the controlling aims of science itself, are continually evolving, in a changing intellectual and social environment. To study in an effective and lifelike manner either the History of Scientific Ideas or the Logic and Methods of Science, we must take this evolutionary process seriously. Otherwise, we shall be in danger, as historians, of concerning ourselves too much with particular discoveries or doctrines or persons, with anticipations and anecdotes. And, as philosophers, we may end by replacing the living science which is our object of study by a formal and frozen abstraction, forgetting to show how the results of these formal enquiries bear on the intellectual and practical business in which working scientists are engaged. A purely chronological history of science and a purely formal philosophy of science thus have the same deficiency: each of them neglects to place the scientific ideas which are in question into their intellectual environment, so as to show what, in that particular context, gave these ideas and investigations their merit.

We are now in a position to return to our opening questions. We asked: 'What merits must a good scientific theory possess? What features enable a theory to score over its rivals?' We gave up the search for any single criterion of merit, such as predictive success, and the time has come to consider some alternative account. Our present enquiry does not, by itself, answer the question, but it does suggest a fresh line of interpretation; and by following up this new line of thought a little way we can usefully sum up the general lessons of our argument.

The ideas of science represent a living and critical tradition. They are passed on from generation to generation, but are modified in the course of transmission. In 1850 (say) Professor Jones teaches physics to his bright young student Smith; and the ideas so transmitted are recognizable ancestors of those which, in 1880, Professor Smith teaches in turn to young Robinson. In each generation, some intellectual variations are perpetuated, and become themselves incorporated into the tradition: this, for the historian, is what constitutes 'progress' in science. Likewise for the philosopher of science: some novel theories deserve to survive at the expense of their rivals and predecessors; and the philosopher must analyse the standards by which such scientific variants are judged and found worthy or wanting. There is no single, simple test of merit, and it is not for the philosopher to impose one on science; nor can a historian justly criticize earlier scientists for not jumping straight to the views of 1960. For progress can be made in science only if men apply their intellects critically to the problems which arise in their own times, in the light of the evidence and the ideas which are then open to consideration.

The common task which accordingly faces historians and philosophers of science has parallels elsewhere—in Darwinian biology. In the evolution of scientific ideas, as in the evolution of species, change results from the selective perpetuation of variants. Between the physics lectures of Professor

Jones in 1850 and those of Professor Smith in 1880 lie thirty years, in which a dozen tentative speculations were considered for every one which survived as a change in the established tradition. For every variant which finds favour and displaces its predecessors, many more are rejected as unsatisfactory. So the question 'What gives scientific ideas merit, and how do they score over their rivals?' can be stated briefly in the Darwinian formula: 'What gives them survival-value?'

This reformulation suggests new questions and possibilities. To begin with, we know from biology how a variation which confers an advantage on one species in one environment may have no merits at all for another species, or even for the same species in a different environment. So, in science, the same theoretical move can have merit in dealing with one group of problems, and yet prove an obstacle to progress in another field or situation. We met this earlier, when we saw how arguments which had merit in the theory of illumination were out of place in gravitation theory; and theoretical patterns which were largely unfruitful in chemistry subsequently bore fruit in genetics.

Again, biological species survive and evolve, not by meeting any single evolutionary demand, but because they alone, from the available variants of earlier forms, have successfully met the multiple demands of the environment. It is easy to think up possible 'advantages' in the abstract: to imagine, for example, how men would benefit from having wings with which they could fly. It is more relevant to calculate the price we should pay for wings: such as the ungainly breast-bone needed to support flight. Only if we do this shall we begin to understand why, in the situation actually existing, surviving species are not even 'better-adapted' than they are.

A parallel issue arises in the logic of science. Considering the various merits of a scientific theory singly and in isolation, we may be enticed by abstract but irrelevant ambitions. Why should we not give actual marks—numerical assessments, that is—to rival scientific theories, grading their merits on a scale? Why should we not build up a theory of confirmation, or calculus of corroboration, with which to demonstrate in numerical form the superiority of one theory to another? This dream lies at the heart of much formalized philosophy of science; yet its hopes of fulfilment are strictly limited. Two rival hypotheses may sometimes be so closely related that their relative merits are positively computable: the 'significance tests' of mathematical statistics do for us what can be done in this way. But this happens only where the really difficult intellectual problems do not arise. As soon as we broaden our view, and consider situations calling for conceptual innovations, where there are several demands to be satisfied, the idea of an 'evidential calculus' for scientific theories becomes unrealizable.

Again, philosophers sometimes assert that a finite set of empirical observations can always be explained in terms of an infinite number of hypotheses. The basis for this remark is the simple observation that through any finite set of points an infinite number of mathematical curves can be constructed. If there were no more to 'explanation' than curve-fitting, this doctrine would have some bearing on scientific practice. In fact, the scientist's problem is very different: in an intellectual situation which presents a variety of demands, his task is—typically—to accommodate some new discovery to his inherited ideas, without needlessly jeopardizing the intellectual gains of his predecessors. This kind of problem has an order of complexity quite different from that of simple curve-fitting: far from his having an infinite number of possibilities to choose between, it may be a stroke of genius for him to imagine even a single one. The scientist might, in fact, retort to the logician as the French painter Courbet is said to have replied to the art-critics by commenting that 'it is a hard enough matter to paint a picture at all, let alone a good picture'. The scientist could here justly reply to formal logicians in similar terms.

The parallel with evolution-theory fits our problem even in unexpected ways. To give one instance only: an inheritable variation sometimes appears in a population first by chance, conferring at that time no particular advantage on its possessors; yet this same variation may subsequently become of extreme value to their descendants as a result of changes in the environment. A feature which, originally, had no merit in this way acquires merit quite unpredictably. The parallel change occurs in science. A classic example of this concerns the notion of 'atomic number'. When the chemical substances were arranged in order and tabulated, Dalton's successors treated 'atomic weight' as the fundamental characteristic of an element. When they had listed the chemical elements in order of atomic weight, it proved convenient to indicate their place in this order, and index numbers were allotted. So, the atomic number of a substance was originally no more than the number in the margin of this list. Had there been twenty or twenty-five elements instead of ninety or more, index letters might have been used instead of index numbers, and the place occupied by a particular substance in Mendelieff's table could have been labelled by its 'atomic letter'.

At the outset, accordingly, atomic number had no theoretical significance. If a scientist had said: 'There must be something 1-ish about hydrogen, 6-ish about carbon, and 8-ish about oxygen—what is it?', his question would have been swept aside as fantastic: as though a man were to notice that *Primula Veris* was species No. 325 in a particular flora, and ask what there was 325-ish about primroses. Subsequent work, however, has given the concept of atomic number quite a different status. With the discovery of isotopes and the development of the Bohr-Rutherford theory, it ceased to be a mere index number and became an actual 'property' of the atom—a much more significant property than atomic weight, from the theoretical point of view. For it became identified with the number of unit-charges of positive electricity on the atomic nucleus, and this is the same in all atoms of a given chemical substance, even where—being isotopes—they have slightly different atomic weight. The cipher in this case became, by an intellectual revolution, the Emperor.

As for the future: how science will develop, and how its ideas and aims will change, can no more be foreseen than the biological future of a species. Men in the future may come to impose fresh demands on the products of their scientific work, and may perhaps even discard as irrelevant some requirements we still maintain. In the nature of the case, one can scarcely do more than state this as a bare possibility: for, in the sphere of creation, to foresee a possibility in any detail is to go half-way to creating it oneself. The kind of change we could most easily imagine would simply complete a process which is already visibly beginning: for instance, theories in physics may soon be taken seriously only if they add to their other merits the additional one of 'programmability'—only, that is, if they lend themselves to machine computation, using modern electronic calculators. And no doubt—one may grant—if this additional demand could always be satisfied, without sacrificing the other merits we already look for in physics, it would certainly be an advantage worth having.

. . . . .

Like all great critical activities, Science has not one, but a number of related aims: it must try to satisfy these as far as possible in harmony, and it is entitled to take on fresh aims. Any activity so varied in scope has, inevitably, a history with many phases: many legitimate enquiries had to be undertaken before the modern tribunal of experimental verification could have its present-day relevance. There is room in the scientific activity today also for men of many talents. Speculative imagination, scrupulous honesty, mathematical command, logical perspicuity, as well as experimental inventiveness and ingenuity: these are all relevant to the manifold aims of Science, in its broadest sense. Here we see the most serious defect in the predictivist account of science: it gives the false impression that the possibilities are closed. Once before, in Hellenistic times, scientists came to see their tasks as restricted to mathematical forecasting: what followed was disastrous. For most of us nowadays the task of understanding Nature is a wider one. Prediction is all very well; but we must make sense of what we predict. The mainspring of science is the conviction that by honest, imaginative enquiry we can build up a system of ideas about Nature which has some legitimate claim to 'reality'. That being so, we can never make less than a three-fold demand of science: its explanatory techniques must be not only (in Copernicus' words) 'consistent with the numerical records'; they must also be acceptable—for the time being, at any rate—as 'absolute' and 'pleasing to the mind'.

Index